COLUMBIA REVIEW
HIGH-YIELD BIOLOGY

COLUMBIA REVIEW

HIGH-YIELD BIOLOGY

Stephen D. Bresnick, M.D.

President and Director
Columbia Review, Inc.
San Francisco, California

Williams & Wilkins
A WAVERLY COMPANY

BALTIMORE • PHILADELPHIA • LONDON • PARIS • BANGKOK
BUENOS AIRES • HONG KONG • MUNICH • SYDNEY • TOKYO • WROCLAW
1996

Editor: Elizabeth A. Nieginski
Managing Editor: Alethea H. Elkins
Development Editor: Melanie Cann
Production Coordinator: Danielle Santucci
Designer: Ashley Pound Design
Cover Designer: Ashley Pound Design
Typesetter: Maryland Composition Co., Inc.
Printer: Port City Press
Binder: Port City Press

Copyright © 1996. The copyright shall be in the name of the Author, except in the case of those illustrations that the Publisher develops and has rendered at its own expense.

351 West Camden Street
Baltimore, Maryland 21201-2436 USA

Rose Tree Corporate Center
1400 North Providence Road
Building II, Suite 5025
Media, Pennsylvania 19063-2043 USA

All rights reserved. This book is protected by copyright. No part of this book may be reproduced in any form or by any means, including photocopying, or utilized by any information storage and retrieval system without written permission from the copyright owner.

Printed in the United States of America

First Edition,

Library of Congress Cataloging-in-Publication Data

Columbia Review high-yield biology / edited by Stephen D. Bresnick.—
 1st ed.
     p.   cm.
   ISBN 0-683-18069-X
   1. Biology—Outlines, syllabi, etc.   2. Biology—Examinations—
Study guides.   I. Bresnick, Stephen D.   II. Columbia Review, Inc.
   [DNLM: 1. Biology—outlines.   2. Biology—examination questions.
QH 315.5 C726 1996]
QH315.5.C64   1996
  574´.02´02—dc20
    DNLM/DLC
    for Library of Congress   96-17351
                                                                 CIP

To purchase additional copies of this book, call our customer service department at **(800) 638-0672** or fax orders to **(800) 447-8438**. For other book services, including chapter reprints and large quantity sales, ask for the Special Sales department.

Canadian customers should call **(800) 268-4178,** or fax **(905) 470-6780.** For all other calls originating outside the United States, please call **(410) 528-4223** or fax us at **(410) 528-8550.**

Visit Williams & Wilkins on the Internet: **http://www.wwilkins.com** or contact our customer service department at **custserv@wwilkins.com.** Williams & Wilkins customer service representatives are available from 8:30 am to 6:00 pm, EST, Monday through Friday, for **either** telephone access.

---

                                                    96 97 98 99
                                               1 2 3 4 5 6 7 8 9 10

Contents

Molecular and Cellular Biology
SECTION I

- **1 ENZYMES** 3
  - I Enzyme Structure and Function 3
    - A Introduction to Enzymes 3
    - B Enzyme Structure 3
    - C Enzyme Function 4
    - D Factors that Affect Enzyme Activity 5
  - II Mechanisms of Enzyme Regulation 6
    - A Allosteric Regulation 6
    - B Feedback Inhibition 7
    - C Multienzyme Complexes 7

- **2 PHOTOSYNTHESIS, CELLULAR METABOLISM, AND CELLULAR RESPIRATION** 9
  - I Introduction 9
    - A Photosynthesis 9
    - B Cellular Respiration 9
  - II Photosynthesis 9
    - A Nutrition 9
    - B Organelles of Photosynthesis 9
    - C Photosynthetic Process 10
  - III Cellular Metabolism 15
    - A Metabolic Pathways 15
    - B Chemical Energy and ATP 16
  - IV Cellular Respiration and Fermentation 17
    - A Overview 17
    - B Glycolysis 17
    - C The Krebs Cycle 19
    - D Electron Transport Chain 19
    - E Anaerobic Pathways 20
    - F Regulation of Respiration 22

- **3 DNA FUNCTIONS AND STRUCTURE** 25
  - I DNA Functions and Structure 25
    - A Functions 25
    - B DNA Structure 25

v

  II DNA Replication 27
   A Overview 27
   B The Mechanism of DNA Replication 27
   C Enzymes Involved in DNA Replication 29

4 GENE EXPRESSION: FROM DNA TO PROTEIN 31
  I Overview 31
   A Deoxyribonucleic Acid (DNA) 31
   B Ribonucleic Acid (RNA) 31
   C Gene Expression 31
   D The Central Dogma 31
  II Transcription 31
   A Mechanism of Transcription 31
   B Regulation of Transcription 33
   C Types of RNA 36
  III Translation 38
   A The Structure and Function of Ribosomes 38
   B The Mechanism of Protein Synthesis 38

5 MICROBIOLOGY 41
  I Viruses 41
   A Viral Structure 41
   B Viral Life Histories 41
  II Prokaryotic Cells 43
   A General Characteristics of Prokaryotes 43
   B Prokaryotic Cell Structure 44
   C Prokaryotic Life Histories 45
  III Fungi 46
   A Characteristics of Fungi 46
   B Fungal Structures 47
   C General Life Histories of Fungi 48

6 THE EUKARYOTIC CELL 51
  I Overview 51
   A All Organisms Other than Bacteria Have Eukaryotic Cells 51
   B Eukaryotes Are More Advanced in an Evolutionary Sense than Prokaryotes 51
   C Cell Structure Can Only Be Studied by Microscopic Observation 51
  II The Eukaryotic Cell: Structures and Functions 51
   A Nucleus 51
   B Membrane-Bound Organelles 52
   C Intracellular Membranes 54
   D Production of a Secretory Product 55
   E Plasma Membrane 56
   F Cytoskeleton 61
  III The Eukaryotic Cell Cycle and Mitosis 63
   A Interphase 63
   B Mitosis 64

7 SPECIALIZED EUKARYOTIC CELLS AND TISSUES ... 67

- I Overview ... 67
  - A Specialization ... 67
  - B Tissues ... 67
- II Neural Cells and Tissues ... 67
  - A Nervous System ... 67
  - B Cells of the Nervous System ... 67
  - C Transmission Along the Neuron ... 69
  - D Transmission Across the Synapse ... 71
- III Contractile Cells and Tissues ... 73
  - A Muscle Tissue ... 73
  - B The Sarcomere: Structure and Mechanism of Contraction ... 75
  - C Regulation of Muscle Contraction ... 76
- IV Epithelial Cells and Tissues ... 78
  - A Structure and Function ... 78
  - B Types ... 78
- V Connective Cells and Tissues ... 79
  - A Structure and Function ... 79
  - B Cartilage and Bone ... 81

SECTION II

Physiology

8 THE NERVOUS SYSTEM ... 85

- I Introduction ... 85
  - A Function ... 85
  - B Structure ... 85
- II Peripheral Nervous System ... 85
  - A Structure ... 85
  - B Function ... 87
- III The Central Nervous System ... 89
  - A The Spinal Cord ... 89
  - B The Brain ... 90
- IV Special Sensory Reception and Processing ... 92
  - A Vision ... 92
  - B Hearing ... 94
  - C Olfaction ... 96
  - D Taste ... 97
  - E Touch ... 97

9 THE ENDOCRINE SYSTEM ... 99

- I Function of the Endocrine System ... 99
  - A Introduction ... 99
  - B Hormones ... 99
- II Major Endocrine Glands ... 101

        A   Pituitary Gland 101
        B   Thyroid 102
        C   Parathyroid Gland 103
        D   Pancreas 103
        E   Adrenal Glands 103

10   THE CIRCULATORY SYSTEM 105

    I   Mechanisms of Circulation 105
        A   Function 105
        B   Structures 105
    II   Mechanisms of Circulation 109
        A   Adult Circulation 109
        B   Fetal Heart Circulation 109
    III Blood Pressure 110
        A   Systolic Blood Pressure 110
        B   Diastolic Blood Pressure 110

11   THE IMMUNE SYSTEM 111

    I   Introduction 111
    II   Types of Immunity 111
        A   Natural Immunity 111
        B   Acquired Immunity 111
    III Cells 112
        A   Lymphocytes 112
        B   Agranulocytes 114
        C   Granulocytes 114
    IV Tissues 114
        A   Bone Marrow 114
        B   Spleen and Lymph Nodes 114

12   THE DIGESTIVE SYSTEM 117

    I   Nutrition 117
        A   Introduction 117
        B   Nutrients 117
    II Gastrointestinal Tract 118
        A   Passage of Food 118
        B   Mouth and Pharynx 120
        C   Esophagus 120
        D   Stomach 120
        E   Small Intestine 121
        F   Colon 122
    III Accessory Organs 122
        A   Liver 122
        B   Pancreas 123

13   THE EXCRETORY SYSTEM 125

    I   The Role of the Excretory System in Body
       Homeostasis 125
    II Mammalian Kidney Structure and Function 125

	A	Gross Anatomy 125
	B	Microscopic Anatomy and Function 126
	C	Formation of Urine 127
	D	Concentration Mechanism 129
	E	Hormonal Influences on the Nephron 130

14 THE MUSCLES AND THE SKELETAL SYSTEM 133

- I Muscle System 133
  - A Functions 133
  - B Structural Organization 133
  - C Types of Muscle 133
  - D Nervous Control 133
- II Skeletal System 134
  - A Function 134
  - B Structure 135
  - C Bone 136
  - D Cartilage 138

15 THE RESPIRATORY SYSTEM 139

- I Function and Basic Anatomy of the Airways and Lungs 139
  - A Function 139
  - B Anatomy 139
- II Hemoglobin, Gas Exchange, and Respiratory Equations 140
  - A Hemoglobin 140
  - B Gas Exchange 142
  - C Equations for Oxygen Loading and Unloading 143
- III Mechanics of Breathing 144
  - A Inhalation and Exhalation 144
  - B Lung Spaces and Pressures 145
- IV Thermoregulation 146
  - A Increases in Lung Ventilation Increase Heat Loss from the Respiratory System 146
  - B Mammals and Birds Control Heat Loss via the Respiratory System to Regulate Body Temperature 146
- V Protective Mechanisms Against Disease and Particulates 146
  - A Upper Respiratory Tract 146
  - B Nasal Cavities 146
  - C Lung 146

16 SKIN 147

- I Structure of the Skin 147
  - A Epidermis 147
  - B Dermis 147
  - C Hypodermis 148
- II Function of the Skin 148

      A   Homeostasis and Osmoregulation 148
      B   Thermoregulation 149
      C   Physical Protection 149

17   THE REPRODUCTIVE SYSTEM 151

    I   Male Reproductive System 151
      A   Genitalia and Gonads 151
      B   Spermatogenesis 152
   II   Female Reproductive System 154
      A   Genitalia and Gonads 154
      B   Oogenesis 155
      C   Menstrual Cycle 156
      D   Pregnancy 158

18   DEVELOPMENT 161

    I   Human Embryology 161
      A   Fertilization 161
      B   Cleavage 163
      C   Blastulation 163
      D   Gastrulation 163
      E   Neurulation 165
   II   Developmental Mechanisms 166
      A   Differentiation 167
      B   Determination 167
      C   Induction 167

SECTION III   Genetics, Evolution, and Botany

19   GENETICS 171

    I   Introduction 171
      A   Basic Definitions 171
      B   Sexual Reproduction 172
   II   Mendel and the Principles of Heredity 172
      A   Mendelian Inheritance 172
      B   Independent Assortment 175
  III   Population Genetics 175
      A   Frequency of Dominant and Recessive Alleles 175
      B   Hardy-Weinberg Principle 176
   IV   Meiosis 177
      A   Process of Meiosis 177
      B   Comparison of Meiosis I and Mitosis 179
      C   The Role of Meiosis in Genetic Variability 179
    V   Mutations 180
      A   Definition and Types 180
      B   Mechanisms of Chromosome Alterations 180

       C   Effects of Mutations on Proteins 181
  VI Sex Linkage 181
       A   Sex Determination 181
       B   Sex-Linked Genes 182
  VII Pedigree Analysis 183
       A   Pedigree and Family History 183
       B   Inheritance Pattern 183
  VIII Gene Mapping 184
       A   Genetic Linkage 184
       B   Mapping Genes 184

20   EVOLUTION 187

  I Natural Selection 187
       A   Charles Darwin 187
       B   Nonadaptive Mechanisms of Population Change .. 187
  II The Concept of Species 188
       A   Definition 188
       B   Reproductive Isolation 188
       C   Speciation 188
  III The Origin of Life 188
       A   Chemical Evolution 188
       B   Earliest Fossil 189
       C   Classification 189
  IV Comparative Anatomy 190
       A   Homology 190
       B   Analogy 190

21   AN OVERVIEW OF BOTANY 191

  I Plant Evolution 191
       A   Protists 191
       B   Four Periods of Plant Evolution 191
  II Plant Anatomy 192
       A   Cellular Structure 192
       B   Roots 192
       C   Stems 194
       D   Leaves 194
       E   Flowers 196
  III Life Cycle 197
       A   Gametophyte 197
       B   Sporophyte 197
  IV Kingdom Plantae 197
       A   Nonvascular Plants 197
       B   Vascular Plants (seedless) 198
       C   Vascular Plants (with seeds) 199

REVIEW QUESTIONS

Section I 201
Section II 213
Section III 232

Preface

High-Yield Biology is one of four books in the *High-Yield College Science Review Series,* published by Williams & Wilkins. In addition to *High-Yield Biology,* the series contains *High-Yield General Chemistry, High-Yield Organic Chemistry,* and *High-Yield Physics.* This series has been designed to make these four important college sciences easier to understand and master. Many students work their way through college courses without really understanding the material they are supposed to be learning. This series is designed to help students improve course grades as well as prepare for post-graduate and pre-professional tests, such as the GRE, MCAT, DAT, PCAT, VET, and OAT. Pre-health majors, biology majors, and even non-science majors will benefit from the books in this series.

High-Yield Biology is an easy-to-read, concise review book of general biology. The book focuses on a conceptual review of basic biology topics, including cellular and molecular biology, physiology, basic genetics, evolution, and botany. The beauty of this book is that it covers an amazing amount of material without being verbose. Many examples, sample problems, tables, and figures enhance the text, written in a narrative outline format. To ensure mastery of the material, over 200 review questions with comprehensive explanations are provided at the end of the book.

About the Author

The author of this series, Dr. Stephen Bresnick, is an expert in helping students understand, review, and retain basic college science material. He is a physician and educator who is active in classroom teaching, scientific research, and writing science review materials for college students. He is currently the Director of **Columbia Review,** a national test preparation company specializing in science and English skill reviews for students interested in entering medical school.

Acknowledgments

The author wishes to thank Drs. William Bresnick, Chris Leptak, and Lauren Yasuda for their contributions. In addition, many thanks to the staff of Williams & Wilkins for their dedication in creating a great, high-yield review book for biology. I especially wish to thank Donna Siegfried, Danielle Santucci, Elizabeth Nieginski, Jane Velker, Tim Satterfield, and Kevin Thibodeau for their expertise and assistance with this important project.

SECTION I

Molecular and Cellular Biology

Enzymes

I. Enzyme Structure and Function

A. Introduction to Enzymes

1. **Every activity in a living cell involves chemical reactions.** In these reactions, the chemical bonds in organic molecules are broken so that new bonds can form.

2. The molecules need an **initial input of energy** from their surroundings for a reaction to start. This energy—the **activation energy (E_A)**—can be thought of as an energy "bump" that must be overcome before a reaction can occur (Figure 1-1A).

3. **Enzymes** are proteins that lower the E_A of reactions (see Figure 1-1B).

   a. Enzymes **reduce E_A** through various mechanisms. For example, enzymes can bring molecules together in the proper orientation to react or can provide a microenvironment conducive to a reaction.

   b. Enzymes act as **biologic catalysts;** they speed up reactions without being changed themselves.

   c. **Without enzymes,** an E_A could not be overcome at the normal temperature of a cell. The chemical reactions would occur so slowly that the cell would die.

4. **Every cell makes many different types of enzymes,** each of which catalyzes a different and highly specific reaction.

   a. Each type of enzyme has a **specific shape** that exclusively binds a set of other molecules.

   b. **Substrates** are the reactants that are acted on in an enzyme-catalyzed (enzymatic) reaction.

B. Enzyme Structure

1. **Enzymes are proteins** (with a few exceptions). Each enzyme has a different, highly precise conformation resulting from several levels of protein structure.

   a. **Primary structure** is the amino acid sequence of the polypeptide chain (or chains) that makes up the enzyme.

   b. **Secondary structure** results from weak chemical bonds (e.g., hydrogen bonds) formed between atoms along the backbone of the polypeptide chain. These are

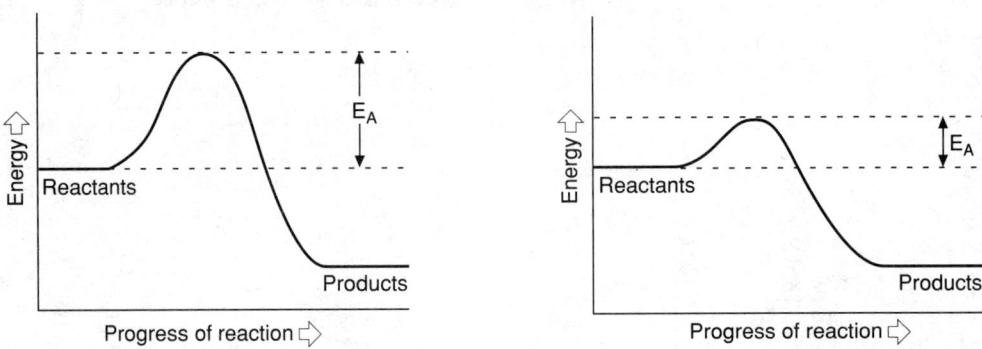

Figure 1-1. **A)** Reaction without enzyme. Note that the reaction requires more energy than the reaction in **B)**, a reaction with enzyme. E_A = activation energy.

        local interactions that result in repetitive three-dimensional patterns (e.g., alpha-helices, beta-pleated sheets).

    c. **Tertiary structure** involves long-distance interactions between amino-acid side chains. These give the protein a highly precise globular shape.

    d. **Quaternary structure** refers to the interaction between two or more different polypeptide subunits of a functional protein.

2. **Active site.** The active site is the restricted area of the enzyme where the substrate (or substrates) binds and where the enzymatic reaction occurs. An active site may be a pocket or a groove in the enzyme molecule.

C. Enzyme Function

1. **Catalytic cycle** (Figure 1-2)

    a. **First step: substrate binds to enzyme.** The substrate or substrates bind to the active site to form the enzyme–substrate complex.

    b. **Second step: induced fit.** Substrate-binding induces a change in the shape of the enzyme so that the substrate fits more snugly in the active site (i.e., induced fit). Induced fit is a reversible change in the enzyme.

    c. **Third step: catalysis.** When the reaction is catalyzed, the substrate or substrates are changed in a specific way, such as by chemical modification, cleavage, or the joining of multiple substrates.

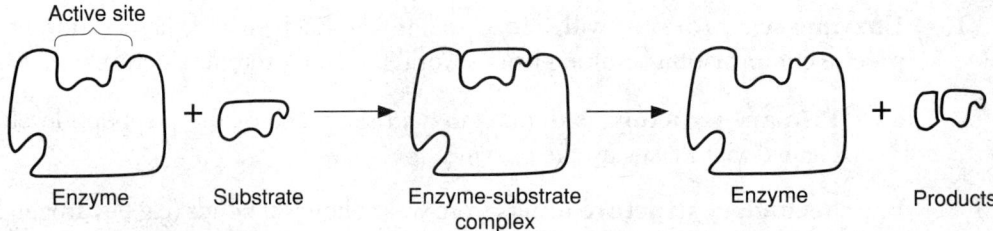

Figure 1-2. The interaction of substrate with enzyme, forming an enzyme–substrate complex. The complex breaks down to reform the enzyme and release the products.

- (1) **Turnover number.** Catalysis occurs so rapidly that a single enzyme molecule can convert more than 1000 substrate molecules per second, which is called the turnover number of an enzyme.
- (2) **Bidirectional.** The same enzyme catalyzes a given reaction in the forward or the reverse direction.
  - d. **Fourth step: products are released.** The products of the reaction are released from the active site, and the enzyme remains in its original form. The enzyme can then leave the active site and be reused with new substrate.
2. **The rate of a reaction** (i.e., the total amount of product per unit time) depends largely on the relative concentrations of substrate and enzyme molecules. As the substrate concentration increases, the rate of the reaction increases because more molecules of enzyme are occupied.
   - a. **Saturation.** At a certain level of substrate, the enzyme becomes saturated (i.e., the active sites of all the enzyme molecules are occupied).
   - b. At the point of saturation, the reaction proceeds at its maximum rate. This rate depends only on how fast each enzyme molecule can convert substrate.

D. Factors that Affect Enzyme Activity

Several factors influence the catalytic activity of enzymes.

1. In many reactions, small **nonprotein substances** are needed for proper enzyme activity. These substances "trigger" a reaction by binding in a specific way to the enzyme molecule.
   - a. **Coenzymes** are organic substances (e.g., vitamins, coenzyme A, biotin, heme).
   - b. **Cofactors** are inorganic substances (e.g., metal atoms of zinc, iron, copper).
   - c. **Prosthetic groups.** Coenzymes or cofactors may bind so tightly that they are effectively part of the protein, at which point they are called prosthetic groups.
   - d. **The holoenzyme** is the protein and nonprotein portions of an enzyme together.
   - e. **The apoenzyme** is the protein portion alone.
2. Each enzyme has **optimal environmental conditions** that favor the most active enzyme conformation.
   - a. **Temperature**
     - (1) **Effect on reactions.** A mild increase in temperature speeds up reactions; molecules move faster and are therefore more likely to interact. A mild decrease in temperature has the opposite effect.
     - (2) **Denaturation.** When a certain temperature is exceeded, chemical bonds are broken, and the enzyme loses its specific shape (i.e., it is denatured). Denaturation is a permanent change that inactivates an enzyme.
   - b. **pH.** An environment that is too acidic or too alkaline can denature enzymes. For most enzymes, the optimum pH is neutral (pH 7). There are exceptions. For example, digestive enzymes in the stomach are active at pH 2.

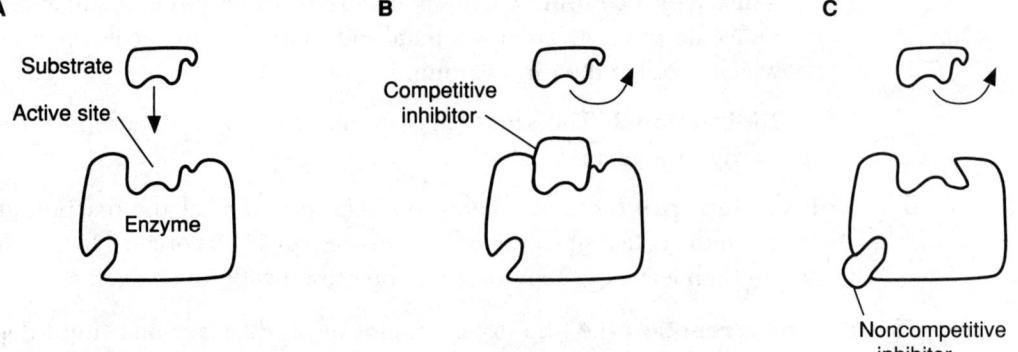

Figure 1-3. **A)** The binding of normal substrate to the active site. **B)** The binding of competitive inhibitor to the active site. **C)** The binding of a noncompetitive inhibitor to a site other than the active site.

3. **Inhibitors** are molecules that bind selectively to enzymes and inhibit enzyme activity. Some inhibitors bind to enzymes reversibly, and others bind irreversibly.

   a. **Competitive inhibitors** resemble the normal substrate and compete with it for binding to the active site of an enzyme (Figure 1-3A). Binding of the inhibitor, therefore, blocks the active site from the substrate (see Figure 1-3B). If the inhibitor is reversible, the effect of competitive inhibition can be overcome by an increase in substrate concentration.

   b. **Noncompetitive inhibitors** bind to a part of an enzyme other than the active site (see Figure 1-3C). Inhibitor binding changes the shape of the active site so that it cannot bind substrate.

II. Mechanisms of Enzyme Regulation

A. Allosteric Regulation

1. An **allosteric site** is a specific site on an enzyme other than the active site. Binding of regulatory molecules to an allosteric site stabilizes one or another alternative conformations of the active site of the enzyme.

   a. **Allosteric binding of an activator** stabilizes an active conformation of the enzyme.

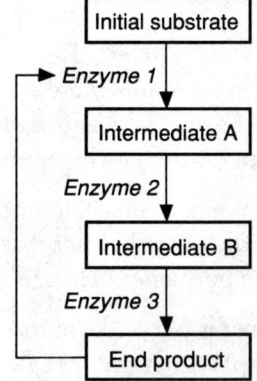

Figure 1-4. The feedback inhibition loop.

        b. **Allosteric binding of an inhibitor** stabilizes an inactive enzyme conformation.

   2. **Allosteric enzymes** are usually protein complexes with multiple subunits, with each subunit having its own active site. Binding of a regulatory molecule to an allosteric site between the subunits can cooperatively affect the entire complex.

B. Feedback Inhibition

During feedback inhibition, a metabolic pathway is regulated by the end product of that pathway (Figure 1-4).

   1. **When the end product is in excess,** it allosterically inhibits the activity of an early enzyme in the pathway. This negative regulation prevents the unnecessary and potentially harmful accumulation of excess intermediates and end products.

   2. **When the end product is no longer in excess,** the enzyme is released from the allosteric inhibition. This results in reactivation of the pathway.

C. Multienzyme Complexes

Several enzymes acting in the same pathway may be assembled together so that the reactions occur efficiently and in the proper sequence (analogous to an assembly line). Multienzyme complexes may be localized in specific cell compartments or organelles.

Photosynthesis, Cellular Metabolism, and Cellular Respiration (2)

I. Introduction

Photosynthesis and cellular respiration can be thought of as opposite, yet interdependent, processes.

A. **Photosynthesis** involves the conversion of light energy, carbon dioxide, and water into glucose, other sugars, and organic compounds; it is the most important mechanism by which oxygen is produced. Oxygen is required for the final stage of cellular respiration.

B. **Cellular respiration** involves the conversion of glucose and other sugars into high-energy phosphate compounds, carbon dioxide, and water. Photosynthesis prevents toxic carbon dioxide levels from accumulating in the atmosphere.

II. Photosynthesis

A. Photosynthesis provides **nutrition** to almost all living things, either directly or indirectly.

1. **Photoautotrophs** use photosynthesis directly to synthesize organic molecules. Photoautotrophs include:

   a. **Plants**

   b. **Multicellular algae** (e.g., pond-living algae, kelp)

   c. **Some unicellular protists** (e.g., *Euglena*)

   d. **Some prokaryotes** (e.g., cyanobacteria)

2. **Heterotrophs** cannot produce their own organic food molecules; they obtain nutrition by eating other organisms or their byproducts. Heterotrophs include:

   a. **Multicellular eukaryotes** (e.g., sponges, insects, fish, amphibians, mammals)

   b. **Fungi,** which feed on decomposing organic matter

   c. **Some prokaryotes** (e.g., unicellular bacteria that act as decomposers)

B. **Organelles of Photosynthesis**

1. **Chloroplasts** are the organelles of photosynthesis of **plants, multicellular algae,** and **protists.**

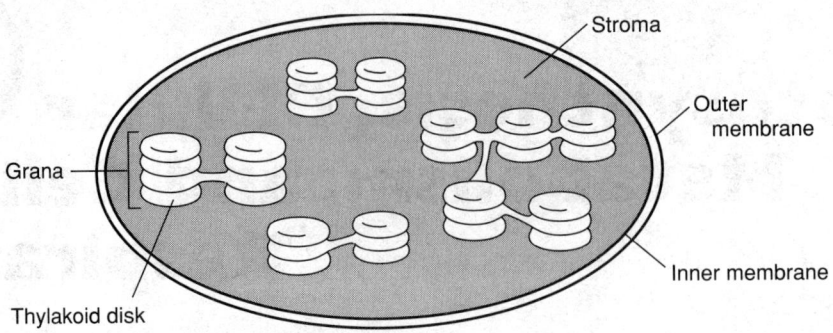

Figure 2-1. Cross-section of a chloroplast.

    a. Chloroplasts contain an **inner membrane**, an **outer membrane**, a **fluid-filled stroma**, and **grana**, stacks of flattened sacs called **thylakoids** that form an extensive membrane system (Figure 2-1).

    b. All green parts of the plant (i.e., the stem and leaves) contain intracellular chloroplasts. In multicellular algae and protists, the chloroplasts are also intracellular.

  2. **Plasma membranes** or **vesicle membranes** are the site of photosynthesis in **photosynthetic prokaryotes**; these organisms lack chloroplasts.

C. Photosynthetic Process

  1. The **photosynthetic reaction** is:

$$6\ CO_2 + 6\ H_2O + \text{light energy} \rightarrow C_6H_{12}O_6 + 6\ O_2$$

  2. **Pigments.** The thylakoid membrane contains pigments that are sensitive to, and absorb, different wavelengths of light energy. All absorbed light energy is ultimately conveyed to chlorophyll *a*.

    a. **Chlorophyll *a*** and **chlorophyll *b*** are composed of a magnesium-containing porphyrin ring and a hydrocarbon tail; the functional group bound to the porphyrin is what differentiates the two. Chlorophyll is **blue-** or **yellow-green.**

    b. **Carotenoids,** which are hydrocarbons, are **yellow** and **orange** pigments.

  3. **Photosystems.** The light-absorbing pigments and some other molecules are arranged into photosystems in the thylakoid membrane. Each photosystem is optimally stimulated by a different spectrum of light.

    a. **Photosystem I** contains chlorophyll *a* molecules that are most sensitive to light at wavelengths of 700 nm (**P_{700}**).

    b. **Photosystem II** contains chlorophyll *a* molecules that are most sensitive to light at wavelengths of 680 nm (**P_{680}**).

  4. **Stages of photosynthesis.** There are two stages of photosynthesis, the light reaction stage and the Calvin cycle (Figure 2-2).

    a. **Light reaction.** The light reaction has two routes for electron flow: cyclic and noncyclic. Both occur in the thylakoid membrane.

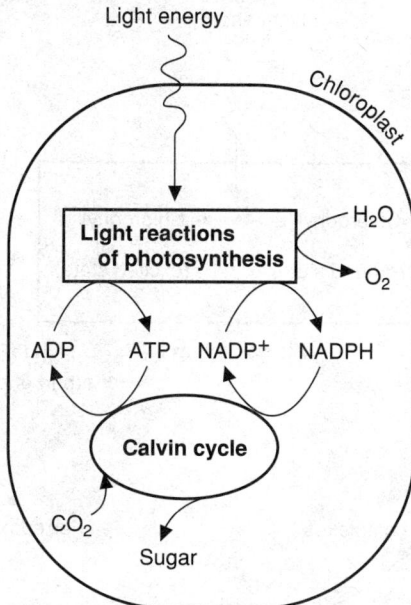

Figure 2-2. An overview of photosynthesis. Adenosine triphosphate (*ATP*) and oxidized nicotinamide adenine dinucleotide phosphate (*NADP⁺*), products of the light reaction stage, are used in the Calvin cycle to produce sugar molecules from carbon dioxide (*CO₂*). *ADP* = adenosine diphosphate; *H₂O* = water; *NADPH* = reduced nicotinamide adenine dinucleotide phosphate; *O₂* = oxygen.

(1) **Cyclic electron flow** (Figure 2-3). Electrons are returned to the chlorophyll in the photosystem.

    (a) Absorption of light energy (photons) by **photosystem I** causes chlorophyll *a* to move from its ground state to an excited state.

    (b) Chlorophyll *a* loses its excited electron to a neighboring molecule, the **primary electron acceptor.**

    (c) The primary electron acceptor supplies electrons to the **electron transport chain.** The chain includes **ferredoxin** (an iron-containing protein), **plastoquinone** (an electron carrier), **cytochrome complexes,** and **plastocyanin** (a copper-containing protein).

    (d) As electrons are being transported from one member of the chain to the next, hydrogen ions are being pumped across the thylakoid membrane into the thylakoid compartment. The **linking of electron transport with hydrogen ion pumping** generates a proton-motive force that powers an enzyme, which in turn phosphorylates adenosine diphosphate (ADP) to adenosine triphosphate (ATP).

    (e) Plastocyanin, the last protein in the electron transport chain, returns the electrons to photosystem I.

(2) **Noncyclic electron flow** (Figure 2-4). Electrons pass from water to oxidized nicotinamide adenine dinucleotide phosphate (NADP⁺); they do not return to the chlorophyll in the photosystem. Either photosystem I or photosystem II pigments can be stimulated.

    (a) **Photosystem I stimulation**

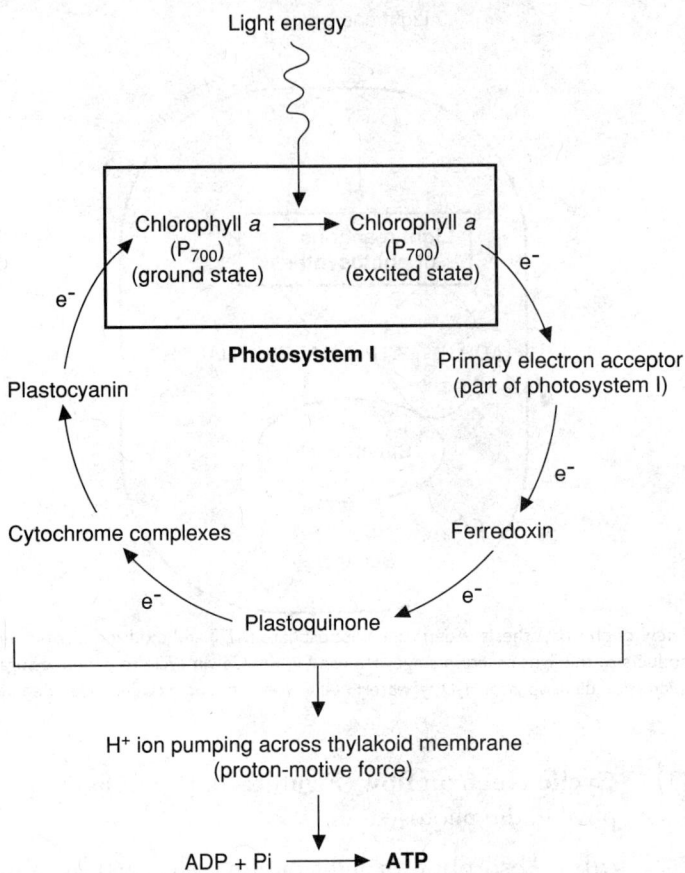

Figure 2-3. Cyclic electron flow of the light reaction. Electrons are returned to the photosystem, and adenosine triphosphate (*ATP*) is generated from adenosine diphosphate (*ADP*). e⁻ = electron; *Pi* = inorganic phosphate.

- (i) Chlorophyll *a* moves from its ground state to an excited state.
- (ii) Electrons are transferred to the primary electron acceptor.
- (iii) From the primary electron acceptor, electrons are transferred to ferredoxin, where $NADP^+$ reductase reduces $NADP^+$ and H^+ to reduced nicotinamide adenine dinucleotide phosphate (NADPH).

(b) **Photosystem II stimulation**

- (i) Light stimulates chlorophyll *a* in the photosystem II complex.
- (ii) The primary electron acceptor of photosystem II pulls electrons from the chlorophyll and transfers them to the same electron transport chain as the one involved in the cyclic reaction.
- (iii) Electrons move down the electron transport chain, losing potential energy. Electron transport is coupled to hydrogen ion pumping across the thylakoid membrane. The proton-motive force generated then drives ATP synthesis.

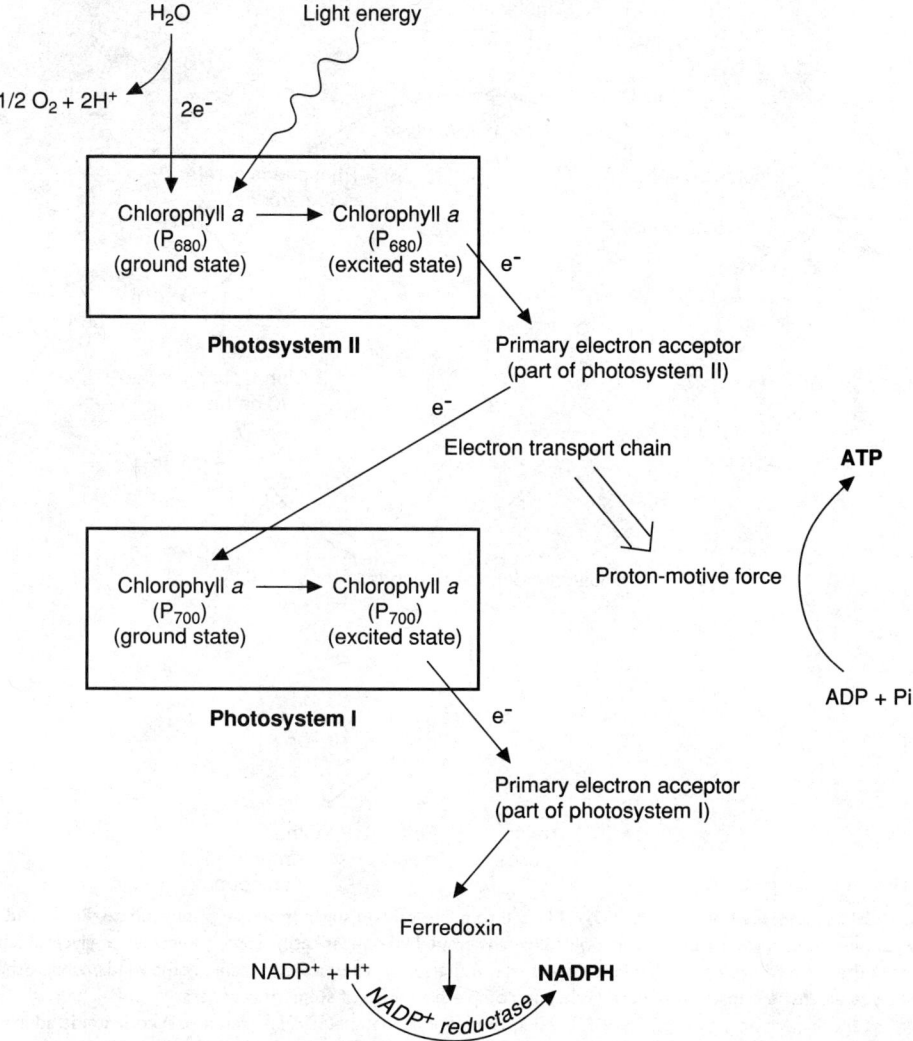

Figure 2-4. Noncyclic electron flow of the light reaction. Electrons are not returned to the photosystem; instead, they are replaced by the hydrolysis of water (H_2O) into oxygen (O_2) and hydrogen (H^+). ADP = adenosine diphosphate; ATP = adenosine triphosphate; e^- = electron; $NADP^+$ = oxidized nicotinamide adenine dinucleotide phosphate; NADPH = reduced nicotinamide adenine dinucleotide phosphate; Pi = inorganic phosphate.

      (iv) Electrons are eventually transferred to photosystem I, where they replace the electrons lost during the reduction of $NADP^+$.

      (v) The electrons lost by photosystem II are replaced by the hydrolysis of water into oxygen gas and hydrogen ions:

$$H_2O \rightarrow \tfrac{1}{2} O_2 + 2H^+ + 2e^-$$

  b. **Calvin cycle** (Figure 2-5). The Calvin cycle, the critical pathway by which sugar molecules are formed from carbon dioxide, occurs in the stroma of the chloroplast.

      (1) To synthesize one molecule of glyceraldehyde phosphate, the cycle must take place three times. $NADP^+$ is used for reducing power, and ATP is used as an energy source. For every three molecules of carbon dioxide

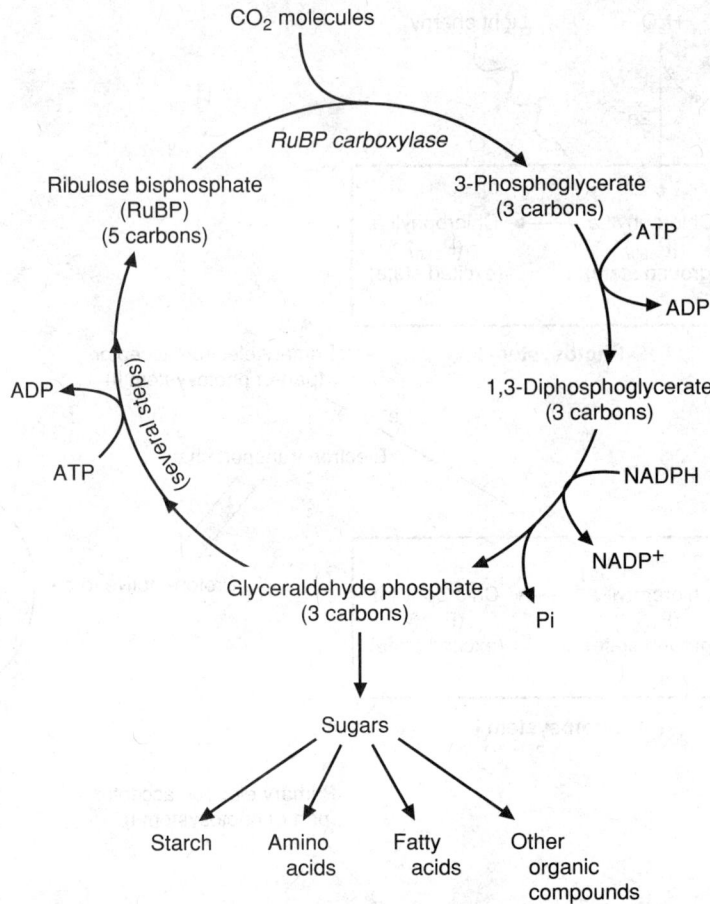

Figure 2-5. Calvin cycle. Carbon dioxide (CO_2) is linked to a five-carbon sugar (ribulose bisphosphate, *RuBP*) and immediately split into two three-carbon sugar molecules (glyceraldehyde phosphate). Some molecules of glyceraldehyde phosphate leave the cycle to provide energy and participate in the synthesis of other organic compounds, while other molecules of glyceraldehyde remain within the cycle and complete a series of steps to regenerate RuBP. The regeneration of RuBP involves adenosine triphosphate (*ATP*). ADP = adenosine diphosphate; NADPH = reduced nicotinamide adenine phosphate dinucleotide; $NADP^+$ = oxidized nicotinamide adenine dinucleotide phosphate; Pi = inorganic phosphate.

that enter the cycle, one molecule of glyceraldehyde phosphate (a three-carbon sugar) is produced, nine ATP molecules are hydrolyzed, and six molecules of NADPH are oxidized.

(a) Each carbon dioxide molecule attaches to ribulose bisphosphate (RuBP), a five-carbon sugar, to form an unstable six-carbon molecule that splits to form two molecules of 3-phosphoglycerate. This step is catalyzed by RuBP carboxylase.

(b) Each 3-phosphoglycerate molecule is supplied with an additional phosphate group by ATP, forming 1,3-diphosphoglycerate.

(c) NADPH reduces 1,3-diphosphoglycerate to glyceraldehyde phosphate.

(2) Some glyceraldehyde phosphate molecules leave the Calvin cycle to supply the plant cell with energy and the building blocks for other organic compounds. Other glyceraldehyde phosphate molecules stay within the cycle to regenerate RuBP, a process that requires ATP.

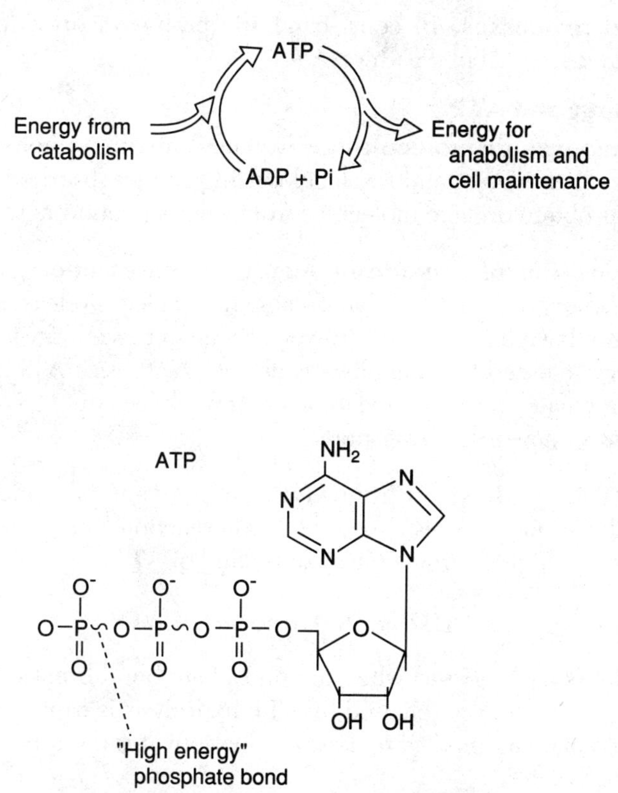

Figure 2-6. **A)** The relationship between catabolism and anabolism, and **B)** the structure of adenosine triphosphate (ATP). Note the position of the key high-energy phosphate bonds. ADP=adenosine diphosphate; Pi=inorganic phosphate.

III. Cellular Metabolism

A. Metabolic Pathways

1. The term **metabolism** refers to the entire set of chemical reactions in a living cell or organism. Metabolism is an essential characteristic of all living organisms.

    a. **Metabolic reactions** occur in "chains" or pathways. In a metabolic pathway, the product of reaction 1 would be a reactant in reaction 2, and so forth. Many pathways are branched because particular molecules enter more than one alternative reaction.

    b. **Enzyme regulation.** Each reaction is catalyzed by a different enzyme. Enzyme regulation controls which pathways are involved in a reaction and to what extent (see Chapter 1).

2. There are two general types of metabolic pathways in living cells (Figure 2-6A).

    a. **Catabolic pathways** result in the decomposition of organic molecules into their simpler components. Catabolism releases the chemical energy stored in the chemical bonds of organic molecules.

    b. **Anabolic pathways** result in the synthesis of organic molecules from their simpler components. Anabolism requires an input of chemical energy, storing the energy in organic molecules.

3. **Coupled processes.** In cells, catabolic pathways provide the chemical energy needed to drive anabolic pathways.

B. **Chemical Energy and ATP**
1. **Obtaining organic molecules.** Only plants and other photosynthetic organisms can synthesize their own organic molecules using energy absorbed from sunlight. All other organisms obtain organic molecules from their surroundings.

2. **Decomposition of organic molecules releases energy.** In all living cells, the chemical energy that is stored in organic molecules is released when these molecules are broken down in catabolic pathways. Some of this energy is lost as heat. The rest of the energy released by catabolism is used to synthesize ATP (see Figure 2-6B). ATP is the immediate energy source for cellular work (i.e., energy-requiring processes such as synthesis, movement, transport).

   a. **ATP is synthesized** from ADP and inorganic phosphate (Pi) with an input of cellular energy. This phosphorylation reaction stores chemical energy in the unstable ("high-energy") phosphate bond of ATP.

   $$ADP + Pi + energy \rightarrow ATP$$

   b. **ATP is hydrolyzed** when the phosphate bond breaks, releasing the stored energy. The energy released by ATP hydrolysis is captured by the transfer of the phosphate to another molecule. This activates the other molecule to do work.

   $$ATP \rightarrow ADP + Pi + energy$$

3. **ATP synthesis** is driven by catabolism of the six-carbon sugar glucose. The energy in glucose is released by either of two catabolic pathways.

   a. **Cellular respiration** is the most prevalent and efficient catabolic pathway. Respiration requires the presence of oxygen (i.e., aerobic conditions).

   b. **Fermentation** is a less efficient pathway and does not require oxygen. Fermentation occurs in many microorganisms, and it occurs under anaerobic conditions in some cells of higher organisms.

4. **Glucose catabolism** involves electron transfer reactions. The stepwise transfer of high-energy electrons from glucose releases energy to drive ATP synthesis.

   a. **Coupled oxidation-reduction reactions** ("redox" reactions) involve the transfer of electrons from one compound to another.

      (1) **Oxidation** occurs when an organic compound loses (donates) electrons. **Reduction** occurs when a compound gains (accepts) electrons.

      (2) In a redox reaction, one molecule donates electrons, and the other accepts electrons. The electron donor is oxidized and the electron acceptor is reduced. Xe^- is oxidized to X, Y is reduced to Ye^-.

$$\underbrace{Xe^- + Y \rightarrow X + Ye^-}_{\text{reduction}}^{\overbrace{}^{\text{oxidation}}}$$

(3) Hydrogen ions (protons) are transferred along with electrons in many redox reactions.

b. **High-energy electrons** are released from glucose and transferred to special molecules called **electron carriers,** which include the coenzymes nicotinamide adenine dinucleotide (NAD^+) and flavin adenine dinucleotide (FAD^+).

$$\overset{\text{oxidation}}{X - H_2 + NAD^+ \rightarrow X + NADH + H^+}\underset{\text{reduction}}{}$$

IV. Cellular Respiration and Fermentation

A. Overview

1. The **overall reaction** for cellular respiration is:

$$C_6H_{12}O_6 + 6O_2 \rightarrow 6CO_2 + 6H_2O + \text{Energy}$$

2. **Respiration**

   a. During respiration, **glucose is completely oxidized to carbon dioxide.** Stepwise redox reactions release a substantial amount of energy for ATP synthesis. Overall, approximately 40% of the energy released from glucose is captured as ATP, and the rest is lost as heat.

   b. In respiration, oxygen is the final electron acceptor. **Oxygen is reduced to water.**

   c. **The respiration pathway includes three stages:** glycolysis, Krebs cycle, and the electron transport chain (ETC).

   d. Most of the ATP production occurs in the final stage of respiration.

   e. Glycolysis occurs in the cytoplasm of the cell, whereas the last two stages occur in the mitochondria.

3. **Fermentation.** During fermentation, glucose is only partially oxidized, and an organic molecule is reduced. Only a small amount of ATP is produced because the Krebs cycle and the ETC do not take place.

B. Glycolysis

1. Glycolysis can be summarized in the following equation:

$$\text{Glucose} + 2NAD^+ + 2ADP + 2Pi \rightarrow 2 \text{ pyruvic acid} + 2NADH + 2H^+ + 2ATP$$

2. **The reactions of the glycolysis pathway** (Figure 2-7A) are catalyzed by enzymes in the cytoplasm.

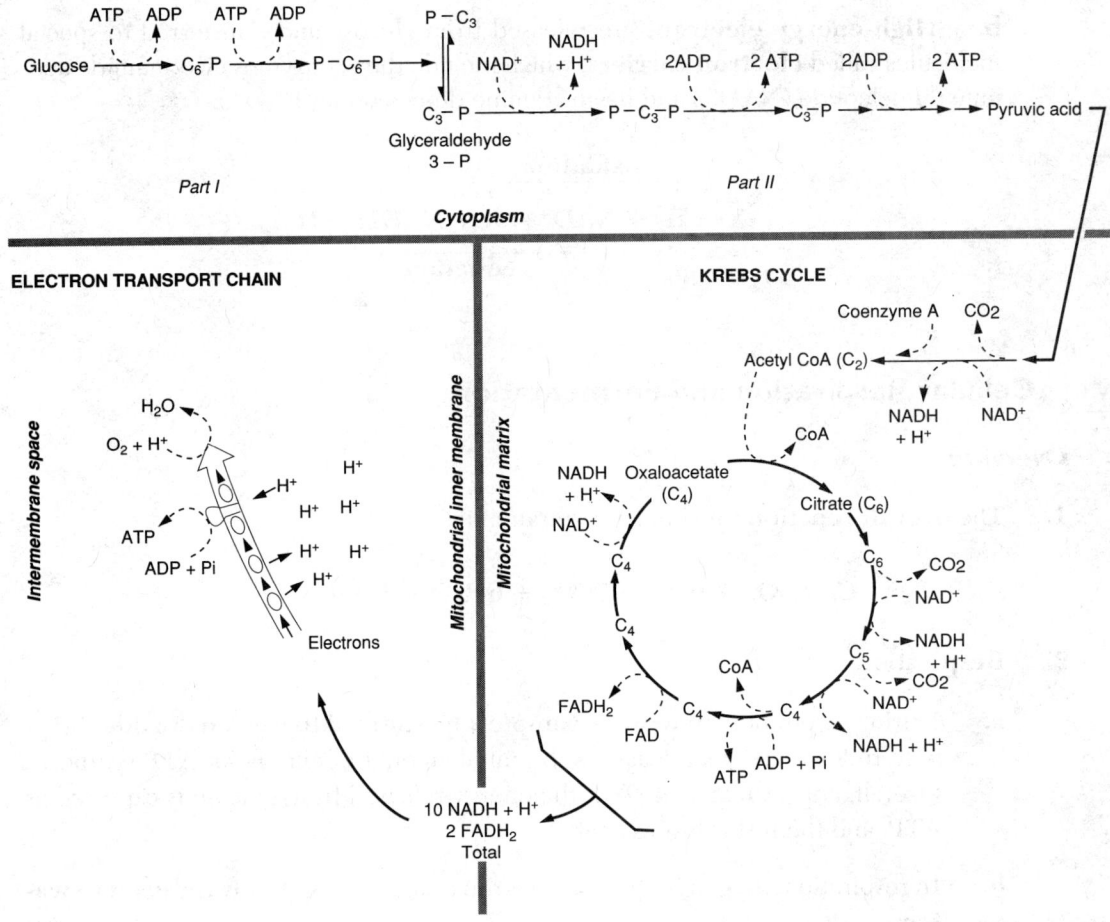

Figure 2-7. An overview of glycolysis, the Krebs cycle, and the electron transport chain (ETC). Note the interrelation of the pathways and the flow of products from one pathway to the next. Also note that in Part II of glycolysis, the pathway is shifted to the right, because only glyceraldehyde-3-phosphate (G3P) directly continues the glycolytic pathway. ADP=adenosine diphosphate; ATP=adenosine triphosphate; FAD=flavin adenine dinucleotide; FADH$_2$=reduced flavin adenine dinucleotide; P$_i$ = inorganic phosphate; NAD$^+$=oxidized nicotinamide adenine dinucleotide; NADH=reduced nicotinamide adenine dinucleotide.

    a.   In the first part of glycolysis, the **preparatory phase,** two molecules of ATP are consumed. Each molecule of glucose is split to form two molecules of a three-carbon compound, glyceraldehyde-3-phosphate (G3P).

    b.   In the second part of glycolysis, the **oxidative phase,** four molecules of ATP are synthesized by the direct transfer of a phosphate group from an intermediate molecule to ADP. This method of ATP synthesis is referred to as **substrate-level phosphorylation:**

$$X - P + ADP \rightarrow X + ATP$$

where X is an organic molecule.

    (1)   G3P is oxidized and two molecules of NAD$^+$ are reduced to NADH.

    (2)   Two molecules of the three-carbon compound pyruvic acid are produced.

3. **The main outcomes of glycolysis** are as follows:

   a. **A net gain of two molecules of ATP per one molecule of glucose**

   b. **Transfer of high-energy electrons from glucose to NADH**

   c. **Partial oxidation of glucose to form pyruvic acid** (most of the chemical energy from glucose is in this organic compound)

      (1) **Pyruvic acid is converted to acetyl coenzyme A (acetyl CoA).**

         (a) In the presence of oxygen, pyruvic acid is transported from the cytoplasm into the mitochondria and converted to a two-carbon compound, **acetyl CoA.**

         (b) One carbon is cleaved from pyruvic acid and released as carbon dioxide, pyruvic acid is oxidized, and NAD^+ is reduced. The resulting two-carbon compound is then attached to coenzyme A to form acetyl CoA.

      (2) **For each molecule of glucose that enters glycolysis, two molecules of acetyl CoA enter the next stage of respiration, the Krebs cycle.**

C. **The Krebs cycle** (see Figure 2-7B) is also known as the citric acid cycle or the tricarboxylic acid (TCA) cycle.

   1. The Krebs cycle occurs in the **inner compartment or matrix of the mitochondria.**

   2. This **series of reactions** can be summarized as follows:

      a. Each molecule of acetyl CoA (a two-carbon compound) enters the Krebs cycle by combining with a four-carbon compound (i.e., oxaloacetate). This results in a six-carbon compound (i.e., citrate).

      b. In subsequent steps, two carbon dioxide molecules are removed from citrate, leaving a four-carbon compound. Stepwise redox reactions oxidize the organic intermediates and reduce three molecules of NAD^+ and one of FAD^+ to NADH and $FADH_2$, respectively.

      c. One molecule of ATP is synthesized by substrate-level phosphorylation.

      d. The four-carbon oxaloacetate is regenerated and is ready to combine with another incoming molecule of acetyl CoA.

   3. The **main outcomes** of the Krebs cycle are as follows:

      a. **Production of two molecules of ATP per glucose molecule**

      b. **Transfer of high-energy electrons to NADH and $FADH_2$**

D. **Electron transport chain** (see Figure 2-7C)

   1. The ETC is composed of a series of electron carriers located in the inner mitochondrial membrane (i.e., cristae). These carriers are mainly proteins with prosthetic groups (e.g., cytochrome with attached iron).

   2. The reduced coenzymes produced by glycolysis and the Krebs cycle (10 NADH and

2 FADH$_2$ per glucose molecule) donate high-energy electrons to the ETC. The electrons are transferred from one carrier to the next in a series of redox reactions.

    a. **The final acceptor is oxygen.** Oxygen has such a high affinity for electrons that it essentially pulls them down the chain.

    b. **Oxygen is reduced to water** when, at the end of the chain, oxygen accepts the electrons combined with hydrogen ions (protons).

3. **Proton pumps and the proton gradient.** As electrons flow down the chain, NADH and FADH$_2$ release protons into the mitochondrial matrix. Some of the inner membrane proteins can pump protons into the intermembrane space (i.e., space between the inner and outer mitochondrial membranes). These **pumps generate an electrochemical concentration gradient** across the inner membrane, which is impermeable to protons. The proton gradient is a source of potential energy.

4. **Energy is released as protons flow down their concentration gradient across the inner membrane.** Protons can cross the membrane only by passing through a membrane-spanning protein complex called ATP synthase. This enzyme uses the released energy to phosphorylate ADP.

    a. **Chemiosmosis** is the term for the coupling of proton flow to ATP synthesis.

    b. During chemiosmosis, three molecules of ATP are generated for each molecule of NADH. Also, two molecules of ATP are generated for each molecule of FADH$_2$, which donates electrons at a later point in the ETC (Table 2-1; Figure 2-8).

5. **Oxidative phosphorylation** is the term used to describe the production of ATP using energy from the redox reactions of the ETC. Most of the ATP output of respiration is produced by oxidative phosphorylation.

E. **Anaerobic Pathways**

1. **Anaerobic pathways** allow glucose catabolism to occur in the absence of oxygen. Table 2-2 compares cellular respiration and fermentation.

TABLE 2-1. Cellular Respiration Summary

Process	Location	Input	Output (per glucose)
Glycolysis	Cytoplasm	Glucose	2 Pyruvic acid, 2 NADH, 2 ATP
Krebs cycle	Mitochondria—inner space (matrix)	2 Pyruvic acid converted to 2 acetyl CoA	2 NADH$^+$ (pyruvic acid conversion step), 6 NADH$^+$, 2 FADH$_2$, 2 ATP
Electron transport chain and oxidative phosphorylation	Mitochondria—inner membrane (cristae)	10 NADH, 2 FADH$_2$	32 ATP (12 H$_2$O)
			*Total: 36 ATP/glucose

Acetyl CoA = acetyl coenzyme A; ATP = adenosine triphosphate; FADH$_2$ = reduced flavin adenine dinucleotide; NADH = reduced nicotinamide adenine dinucleotide.

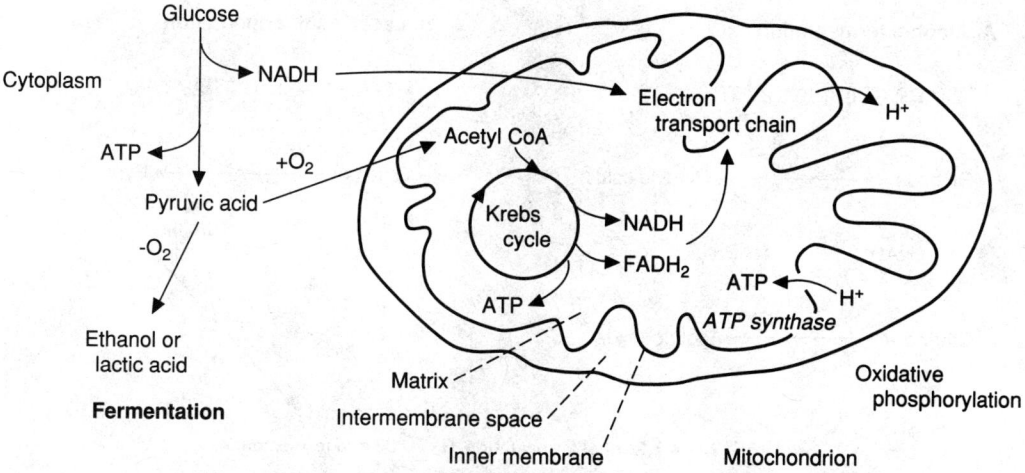
Figure 2-8. Summary of respiration and fermentation.

2. Certain microorganisms are capable of **anaerobic respiration,** in which an inorganic compound is the final electron acceptor instead of oxygen. (For example, some bacteria reduce sulfates or nitrates in soil.)

3. **Fermentation** is an anaerobic pathway that occurs in many different microorganisms as well as in certain cell types in higher organisms (e.g., human muscle cells). The entire pathway occurs in the cytoplasm, and there is no ETC.

   a. **Fermentation pathways start with glycolysis, followed by production of fermentation waste products** (Figure 2-9). This process partially oxidizes glucose to an organic molecule, which is then reduced by NADH. Most of the energy in glucose remains in an organic molecule.

   b. A total of two molecules of ATP per one molecule of glucose are produced by **substrate-level phosphorylation** during glycolysis.

4. The various **types of fermentation pathways** differ in the waste products produced (see Figure 2-4).

   a. In **alcohol fermentation,** pyruvic acid releases carbon dioxide and forms a two-carbon compound, **acetaldehyde.** Acetaldehyde is reduced to form **ethanol,** and NADH is oxidized to regenerate NAD^+.

TABLE 2-2. Comparison of Respiration and Fermentation

	Cellular Respiration	Fermentation
Growth conditions	Aerobic	Aerobic or anaerobic
Electron transport chain	Yes	No
Final electron and hydrogen acceptor	Oxygen	Organic (e.g., acetaldehyde or pyruvic acid)
Method of making ATP	Mostly oxidative phosphorylation; some substrate-level phosphorylation	Substrate-level phosphorylation

ATP = adenosine triphosphate.

SECTION I • PHOTOSYNTHESIS, CELLULAR METABOLISM, AND CELLULAR RESPIRATION

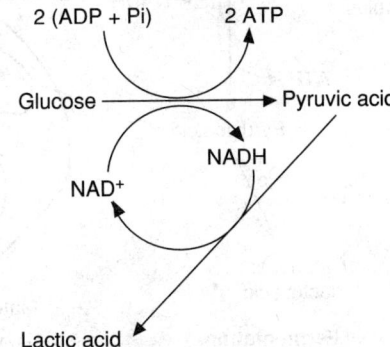

Figure 2-9. **A)** Alcohol fermentation. **B)** Lactic acid fermentation.

      b.   In **lactic acid fermentation,** pyruvic acid is reduced to lactic acid, and NADH is oxidized. Some cells, such as human muscle cells, can switch from cellular respiration to lactic acid fermentation when oxygen is depleted.

F. Regulation of Respiration

    1.   **Respiration is interconnected with the metabolism of all the organic molecules in cells** (Figure 2-10A).

          a.   Carbohydrates, proteins, and fats from food can enter respiration at different points. For example, complex carbohydrates are converted into glucose,

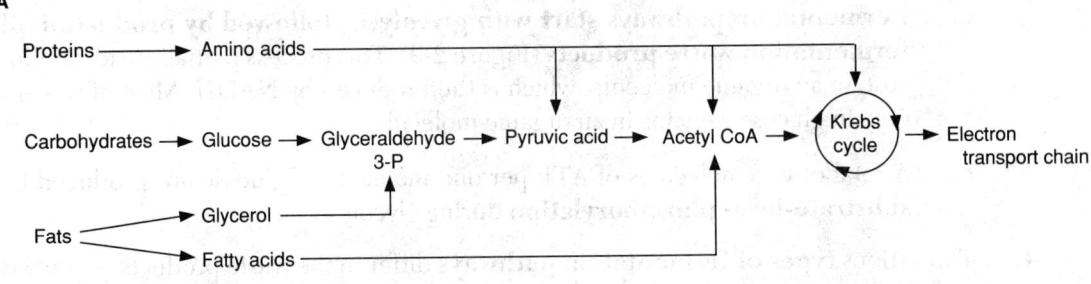

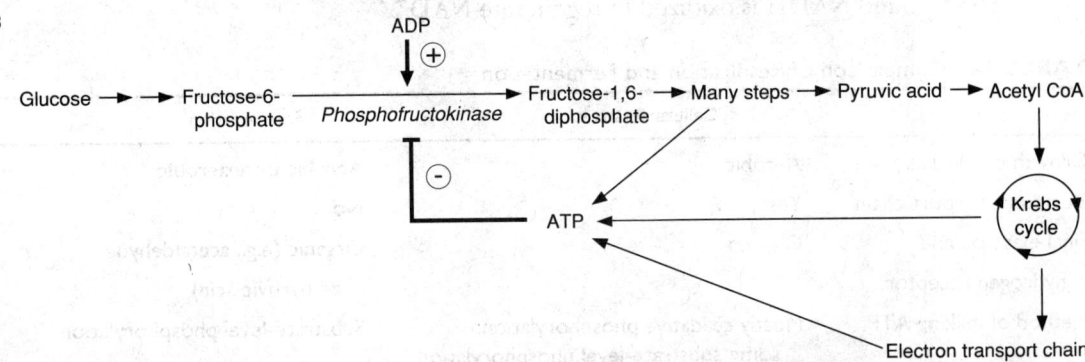

Figure 2-10. **A)** The metabolism of organic molecules and **B)** an example of metabolic regulation by phosphofructokinase, the key rate-limiting step of glycolysis.

proteins are converted into pyruvic acid or Krebs cycle intermediates, and fats are converted into glycolytic intermediates or acetyl.

    b. Conversely, organic intermediates can leave respiration to enter synthesis pathways (anabolism). In this way, carbohydrates can be converted to fats, fats can be converted to proteins, and so forth.

2. The **rate of respiration** is controlled by the enzymes that catalyze the various reactions along the pathway. Many of these are allosteric enzymes with sites for binding activators and inhibitors. **One of the main rate-limiting reactions is the third step of glycolysis, which is catalyzed by the enzyme phosphofructokinase** (see Figure 2-10B). This step is sometimes referred to as the "pacemaker" of respiration.

    a. Phosphofructokinase is inhibited by ATP and is activated by ADP. Thus, when ATP is high, respiration slows down.

    b. When ATP is depleted, the ADP concentration increases, and the enzyme is reactivated. This accelerates the rate of respiration.

DNA Functions and Structure

I. DNA Functions and Structure

A. Functions

1. **Deoxyribonucleic acid (DNA)** is the heritable (genetic) material in all cells. DNA is precisely replicated during each cell generation. When the cell divides, an identical copy of the parental DNA is distributed to each daughter cell. Thus, DNA provides instructions for all future generations of single cells and entire multicellular organisms.

2. **DNA controls the activities of cells** by specifying the synthesis of enzymes and other proteins. Proteins are the class of molecules with the greatest diversity of essential cellular functions; they catalyze and regulate metabolic reactions, provide raw materials for cell structures, allow movement, interact with the environment and other cells, and control growth and cell division.

3. **A gene is the unit of information in DNA.** Each gene specifies the amino acid sequence of a particular protein. Thousands to millions of different genes are needed to make all the necessary proteins in a single cell.

B. DNA Structure

1. DNA belongs to a class of organic molecules called **nucleic acids.** The subunits of nucleic acids are called **nucleotides** (Figure 3-1A). Each nucleotide is composed of the following:

    a. **Deoxyribose,** which is a five-carbon sugar

    b. **A phosphate group,** which is attached to the 5′ carbon of deoxyribose

    c. **A nitrogenous base,** of which there are four different types—two purines and two pyrimidines (see Figure 3-1B).

2. **Nucleotides are linked together by phosphodiester bonds** between the phosphate of one nucleotide and the sugar of the next nucleotide (Figure 3-2A). Alternating sugars and phosphates make up the long "backbone" of the nucleotide chain, or polynucleotide. The bases extend from the 1′ carbon of each sugar in the chain.

3. The two ends of a polynucleotide differ from each other, making it a **polar molecule.** The ends are designated according to the numbered carbon in the sugar. A phosphate group is at the 5′ end, and a hydroxyl group is at the 3′ end.

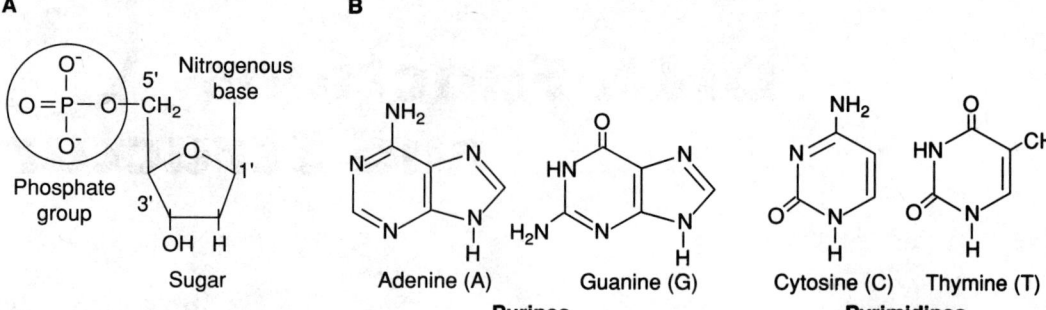

Figure 3-1. **A)** The structure of a nucleotide. **B)** The structure of the nitrogenous bases—the purines and pyrimidines. The structure of uracil, a pyrimidine found in RNA, is not shown. Note the presence of a hydroxyl group on the 3′ carbon of the nucleotide.

4. **The DNA double helix** (see Figure 3-2B)

   a. The structure of DNA was deduced in the early 1950s by James Watson and Francis Crick. They built a model of DNA using information from experiments by Rosalind Franklin and other scientists.

   b. DNA is a **double-stranded molecule,** or duplex, composed of two paired polynucleotides. The pairing results from specific interactions between the bases in each DNA strand. The structure is like a rope ladder in which the sugar-phosphate backbones form the "sideropes" and the pairs of bases form rigid "rungs."

   c. The DNA molecule twists to form a **helix** with 10 bases per helical turn.

   d. The two DNA strands are paired with opposite 5′-to-3′ polarity. In other words, the two strands are **antiparallel.** The 5′ end of one strand is paired with the 3′ end of the other, and they run in opposite directions.

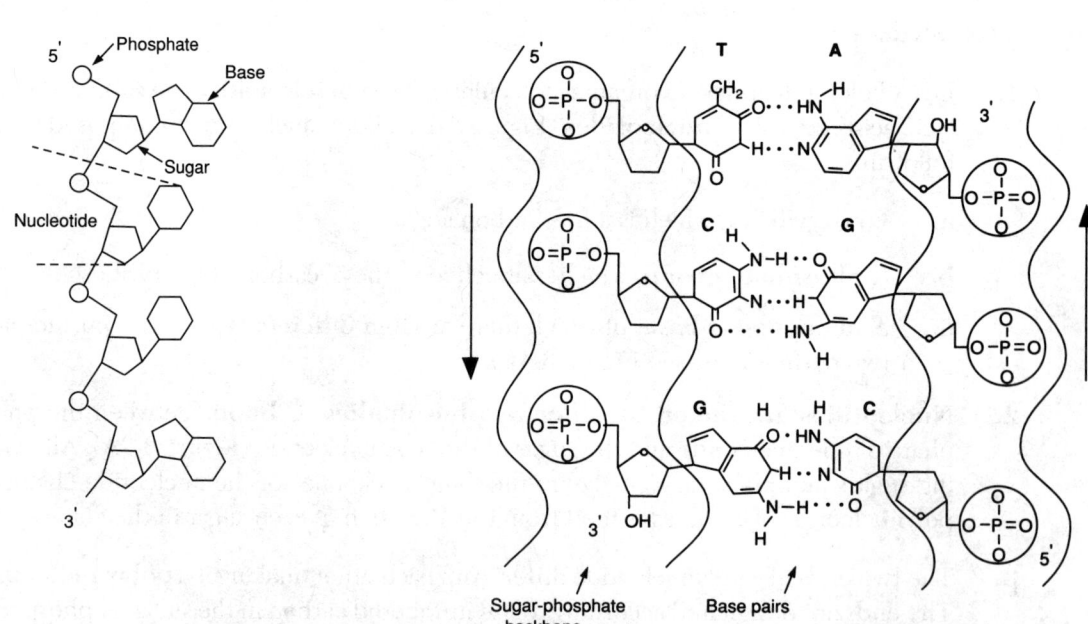

Figure 3-2. **A)** The backbone of DNA is comprised of sugars and phosphates. **B)** Details of DNA at the molecular level.

5. The two strands of a DNA molecule are paired by **complementary base pairing.** Each nucleotide in one strand pairs with a specific (complementary) nucleotide in the other strand.

   a. **Purines and pyrimidines.** Purine bases (i.e., adenine, guanine) are larger and always pair with pyrimidines (i.e., thymine, cytosine), which are smaller. Specifically, **adenine (A) always pairs with thymine (T), and guanine (G) always pairs with cytosine (C).**

   b. **Complementary base pairing** results from the formation of hydrogen bonds (weak, noncovalent bonds) between the specific pairs. Each A–T pair forms two hydrogen bonds, whereas each G–C pair forms three hydrogen bonds. A DNA molecule is stabilized by the large number of hydrogen bonds along its length.

6. The variation in DNA is found in the linear sequence of base pairs along the lengths of the molecules. Amazing sequence diversity is generated in DNA molecules thousands to millions of base pairs (bp) long.

II. DNA Replication

A. Overview

1. **Templates.** One key feature of DNA is its ability to be replicated, so that identical copies of the genes can be passed on to the next generation. Each DNA molecule provides its own templates for replication.

   a. **Each strand serves as a template** for the synthesis of a new complementary DNA strand according to the rules of specific base pairing.

   b. An enzyme called **DNA polymerase** synthesizes DNA by linking each newly paired nucleotide to the 3' end of a growing DNA strand.

   c. **Semiconservative replication.** In each daughter DNA molecule, one parental (old) DNA strand is conserved, and it is paired with one newly synthesized strand. This pattern of replication is referred to as semiconservative (Figure 3-3). In contrast, a **conservative pattern** would result in one intact parental DNA molecule and one completely new DNA molecule. DNA replication has been found to be semiconservative.

   d. **DNA replication is extremely fast and accurate.** Approximately 500 nucleotides are replicated per second, and there is approximately one mistake in 1 billion bp. This is an amazing feat, considering the size of a eukaryotic genome (i.e., approximately 3 billion bp of DNA) within a single cell nucleus.

B. The mechanism of DNA replication was initially observed in the bacteria *Escherichia coli*. A similar process occurs in eukaryotes.

1. **Initiation.** Replication is initiated when specific proteins recognize and bind to special sites on a chromosome called **replication origins.** There are multiple origins on each linear eukaryotic chromosome, but only one origin on the single circular *E. coli* chromosome.

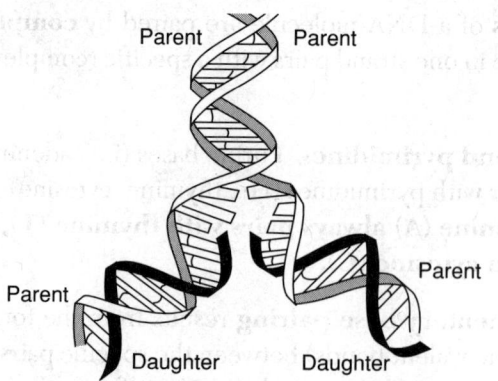

Figure 3-3. The semiconservative replication of DNA.

2. **Processing.** Replication proceeds at structures called **replication forks.** Each fork moves in one direction as more parental sequences are made available for replication. Several processes go on at the replication fork (Figure 3-4).

   a. **5′ to 3′ direction.** Base pairing with a parental template strand determines the sequence of nucleotides in each new DNA strand. The complementary nucleotide is linked to the 3′ end of the new strand by a molecule of DNA polymerase, which moves along the template. Thus, a new DNA strand is always synthesized in the 5′ to 3′ direction.

   b. **Continuous elongation.** At the replication fork, one new strand is elongated continuously in the leading 5′ to 3′ direction toward the moving fork. This is called **leading strand synthesis.**

   c. **Lagging strand synthesis** describes the mechanism of discontinuous elongation, which occurs in the nonleading template strand. Because one template strand is antiparallel, one new strand must be formed in an overall 3′ to 5′ direction, which is not the leading direction.

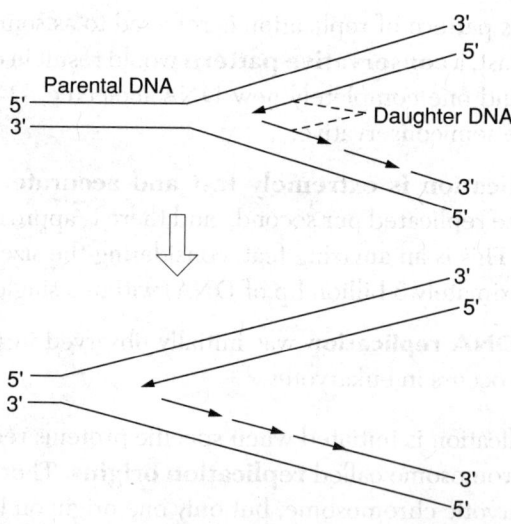

Figure 3-4. The replication of DNA occurring at replication forks.

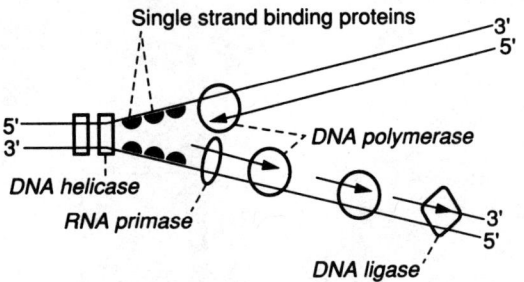

Figure 3-5. The enzymes involved at the replication fork.

- (1) **Okazaki fragments,** which are short DNA fragments, are synthesized in a 5′ to 3′ direction, which is away from the replication fork.
- (2) As the fork moves, additional Okazaki fragments are added in a 3′ to 5′ direction, which is toward the moving replication fork, so that overall growth is in this direction.

d. Different molecules of DNA polymerase synthesize the leading and lagging strands at the replication fork.

C. Enzymes Involved in DNA Replication

1. **Replication complex.** The replication process involves many different enzymes. A replication complex is formed when several of the different enzymes join with DNA polymerase.

2. **Enzymes at the replication fork** (Figure 3-5)

    a. **DNA helicase** unwinds the two parental DNA strands in front of the replication complex.

    b. **Single-strand binding proteins** bind to the separated parental strands to keep them from reannealing (i.e., repairing).

    c. **RNA primase** starts de novo synthesis of each new Okazaki fragment of the lagging strand. The primase synthesizes a short RNA primer complementary to the newly exposed DNA template. The RNA primer provides the 3′ end that DNA polymerase needs to start adding DNA nucleotides.

3. An **exonuclease** removes the RNA primer between fragments.

4. **DNA ligase** joins the Okazaki fragments together to elongate the new continuous DNA strand.

Gene Expression: From DNA to Protein

I. Overview

A. Deoxyribonucleic Acid (DNA)

The information in a gene is specified by the linear sequence of nucleotides in the DNA. A single gene may be hundreds to hundreds of thousands of nucleotides in length.

B. Ribonucleic Acid (RNA)

DNA does not build a protein directly; it sends instructions to the protein synthesis "machinery" in the form of RNA.

1. **Transcription** is the process by which RNA is copied from the DNA. After the DNA is transcribed in the nucleus, the RNA moves into the cytoplasm.

2. **Translation** is the process by which the information in RNA is used to synthesize proteins. Translation occurs in the cytoplasm.

C. Gene Expression

A gene undergoing transcription and translation is said to be expressed (i.e., the information in DNA is used to make a protein).

D. The central dogma of molecular biology is the scheme DNA → RNA → protein.

II. Transcription

A. Mechanism of Transcription

1. **RNA structure.** RNA is a nucleic acid with a structure similar to that of DNA. However, there are several important differences (Table 4-1).

    a. In RNA, a **hydroxyl group** instead of a hydrogen molecule is attached to the 2′ carbon of the five-carbon sugar.

    b. RNA contains the nitrogenous base **uracil (U)** instead of thymine. Uracil pairs with adenine.

    c. RNA molecules are **single-stranded** rather than double-stranded **polynucleotides**.

2. **Transcription of a gene** involves synthesis of a complementary strand of RNA from a DNA template strand (Figure 4-1).

    a. **RNA polymerase** is the enzyme responsible for RNA synthesis.

TABLE 4-1. Comparison of Deoxyribonucleic Acid (DNA) and Ribonucleic Acid (RNA)

Characteristic	DNA	RNA
Sugar	Deoxyribose	Ribose
Bases	A, G, C, T	A, G, C, U
Structure	Double-stranded	Single-stranded
Synthesis process	Replicated from DNA template	Transcribed from DNA template
Functions	Storage and inheritance of information for proteins	Transmittance of DNA information to cytoplasm for translation

A = adenine; C = cytosine; G = guanine; T = thymine; U = uracil.

    b. **Promoter sequence.** Transcription starts when RNA polymerase binds to a specific DNA sequence found near the start of the gene. This transcription start site is called the promoter. Only one strand of the DNA duplex contains the promoter sequence, so only that strand is transcribed.

    c. **Processing.** Transcription proceeds as RNA polymerase moves along the DNA template, linking complementary RNA nucleotides into a growing chain. The only change in the base pairing rules (compared with DNA replication) is that uracil is incorporated into the RNA opposite adenine in the DNA template.

    d. **Elongation.** As in DNA synthesis, the new RNA strand is elongated in the 5′ to 3′ direction. The "start" of a gene (the promoter) is therefore called the 5′ end of the gene. Accordingly, the end of a gene is its 3′ end.

    e. **The DNA duplex "opens up"** during active transcription, which allows many molecules of the RNA polymerase to transcribe the gene at the same time.

    f. **Termination sequence.** Transcription stops at a termination sequence in the DNA. Here, the RNA polymerase leaves the DNA template and releases the newly synthesized RNA.

3. **Coding and noncoding regions.** An entire gene is transcribed into one continuous RNA molecule. However, most eukaryotic genes contain stretches of DNA that do not code for amino acids. The protein-coding regions of a gene are called **exons** (they are expressed as protein), and the noncoding regions are called **introns** (intervening sequences).

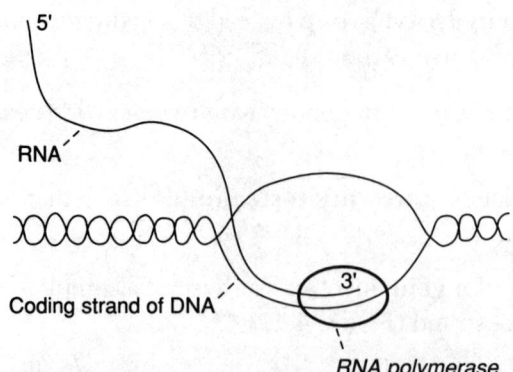

Figure 4-1. The transcription of ribonucleic acid (RNA) from a deoxyribonucleic acid (DNA) template.

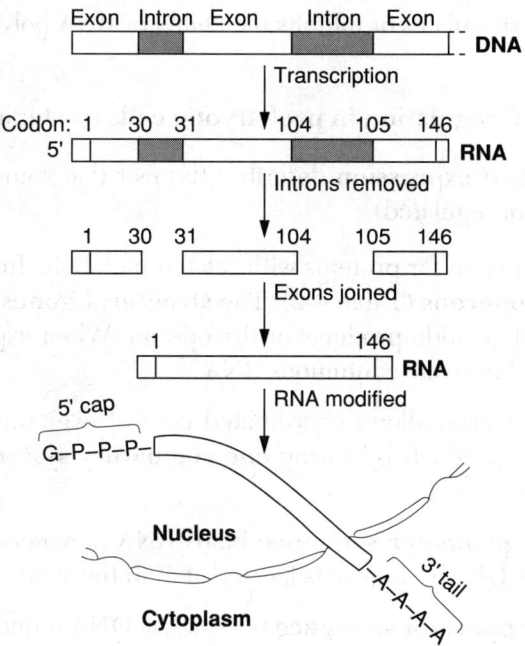

Figure 4-2. Posttranscriptional modification of RNA. Note the RNA splicing and the placement of a "cap" and a polyadenylate (poly A) "tail."

- a. **RNA splicing.** While a newly synthesized RNA molecule is still in the nucleus, the introns are removed (spliced), and the exons are joined together (Figure 4-2). This process results in a continuous protein-coding sequence.

- b. **5′ Cap and poly A tail.** The RNA is also modified at its 5′ and 3′ ends to protect it from degradation. These modifications include a 5′ cap, which is modified guanosine triphosphate (GTP), and a 3′ polyadenylate (poly A) tail.

4. **Nontranscribed regions.** Genes make up only a small fraction of the DNA in a eukaryotic cell. The genes are often separated by large regions of noncoding DNA that are never transcribed.

B. Regulation of Transcription

1. Transcription is a highly regulated process. Many of the genes in the genome of an organism are not expressed in all cells or at all times.

    - a. **Unicellular organisms** are short-lived and vulnerable to their surroundings; they must reproduce rapidly to ensure population survival. In these cells, gene regulation avoids producing unneeded gene products (which would be a waste of energy) and allows rapid adaptation to changes in the surroundings.

    - b. In **multicellular organisms,** many genes are expressed only at specific developmental stages and only in specific cells. This **differential gene expression** produces the many specialized cell types of the tissues, organs, and organ systems that make up the entire organism.

2. Transcription is generally controlled by the **binding of regulatory proteins to specific regulatory DNA sequences** located near or within a gene. This molecular in-

teraction either stimulates or inhibits the ability of RNA polymerase to transcribe the gene.

3. **Transcriptional regulation in prokaryotic cells** (bacteria)

   a. **Constitutive expression** describes the fact that some genes are always transcribed (not regulated).

   b. Genes that code for proteins with related metabolic functions are clustered together in **operons** (Table 4-2). The **structural genes** are the genes that code for the polypeptide products of the operon. When expressed, these genes are transcribed into one continuous RNA.

      (1) An operon allows coordinated control over the expression of multiple gene products by having **one regulatory system** for all of its structural genes.

      (2) The **promoter sequence** binds RNA polymerase to start transcription. This DNA sequence is located at 5′ to the structural genes.

      (3) The **operator sequence** is a specific DNA sequence located between the promoter and the structural genes. A regulatory protein called the **repressor** can bind to the operator and block RNA polymerase from transcribing the genes. The repressor–operator interaction is an "on–off switch" that controls whether the operon is transcribed.

   c. A **repressible operon** is turned off (repressed) by activation of the repressor. An example is the **tryptophan (Trp) operon,** which codes for enzymes in the tryptophan synthesis pathway. Tryptophan serves as the corepressor of the Trp operon: when the amino acid tryptophan begins to accumulate, it binds to and activates the repressor. This turns off the operon, thereby preventing synthesis of excess tryptophan.

   d. An **inducible operon** is turned on (induced) by inactivation of the repressor. An example is the **Lac operon** (Figure 4-3), which codes for enzymes in the lactose catabolism pathway (e.g., β-galactosidase). When lactose is present, it binds to and inactivates the repressor. This turns on the operon, thereby allowing breakdown of the nutrient lactose.

4. The examples in Figure 4-3 illustrate **negative control,** which involves a **repressor protein** that turns transcription off. **Positive control** involves an **activator protein,** which stimulates transcription. An example from the Lac operon is the activity of the catabolite activator protein (CAP). CAP stimulates transcription by interacting with RNA polymerase at the promoter region.

TABLE 4-2. Operon Summary

Operon Function	Example	Transcription	Control
Synthesis of a product	Trp operon	On unless the product is in excess	Repressible
Catabolism (break down) of a nutrient	Lac operon	Off unless nutrient is available	Inducible

Lac = lactose; *Trp* = tryptophan.

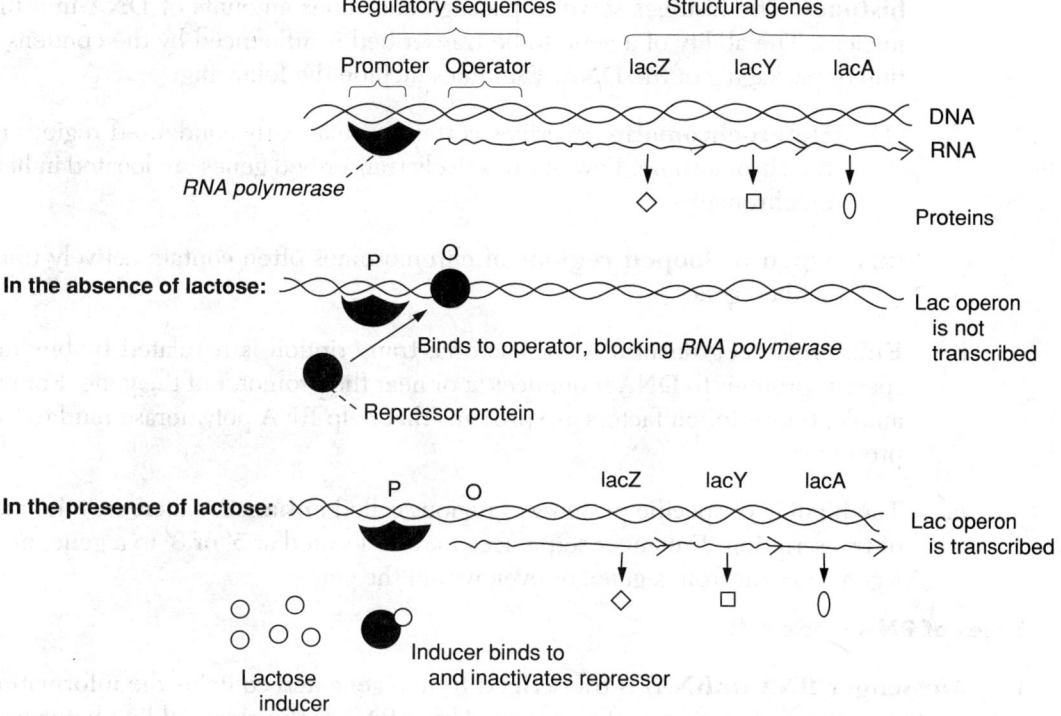

Figure 4-3. The lactose (Lac) operon is a classic example of transcriptional regulation.

    a. CAP is activated when lactose is needed as an energy source because glucose is unavailable.

    b. CAP is activated by interacting with **cyclic adenosine monophosphate (cAMP),** which increases in concentration as glucose levels decrease.

5. **Transcriptional regulation in eukaryotic cells** (Table 4-3)

    a. **Packaging of DNA.** Transcriptional activity is influenced by the organization of the DNA in the chromosomes. Each unreplicated chromosome is composed of a single, linear DNA molecule associated with special proteins called

TABLE 4-3. Comparison of Gene Regulation in Prokaryotes and Eukaryotes

	Prokaryotes	Eukaryotes
General emphasis	Economy, efficiency (rapid growth, short life span)	Cell specificity and cooperation (multicellular organisms)
Gene organization	Approximately 2500 genes; no introns; few DNA-associated proteins; genes cotranscribed in operons	Approximately 100,000 genes, introns, and exons. DNA packaged with proteins; genes transcribed separately
Regulatory mechanisms	Genes have shared regulatory regions for rapid, coordinated control (repression and induction mechanisms)	Many levels of regulation (e.g., chromosome packaging, transcription factors, mRNA processing, translation, protein processing)
Transcription and translation	Simultaneous; both occur in the cytoplasm	Separated in time and space
mRNA	Rapidly degraded	Controlled stability

histones. The histones serve to package enormous amounts of DNA into the nucleus. The ability of a gene to be transcribed is influenced by the condensation or packaging of the DNA. Examples include the following:

(1) **Heterochromatin** describes certain permanently condensed regions of the chromosome. Few or no actively transcribed genes are located in heterochromatin.

(2) **Open or looped regions** of chromosomes often contain actively transcribed genes.

b. **Eukaryotic regulation.** As in bacteria, transcription is regulated by binding specific proteins to DNA sequences at or near the promoter of the gene. For example, transcription factors are proteins that help RNA polymerase bind to the promoter.

c. The binding of specific proteins to regions called **enhancers** increases the rate of transcription. Enhancer sequences may be located at 5′ or 3′ to a gene, near a gene, distant from a gene, or even within the gene.

C. Types of RNA (Table 4-4)

1. **Messenger RNA (mRNA)** is transcribed from a gene and contains the information for the amino acid sequence of a protein. The mRNA is the essential link between a gene and its protein product. After being synthesized in the nucleus, mRNA molecules move out into the cytoplasm where they are translated by the protein synthesis machinery.

   a. **Genetic code** (Table 4-5). The mRNA specifies a sequence of amino acids according to the genetic code. In this code, each continuous set of three nucleotides specifies one amino acid.

   (1) Each nucleotide triplet is called a **codon.** For example, the following mRNA sequence:

   $$\overbrace{AUG}\;\overbrace{CCA}\;\overbrace{GGC}\;\overbrace{AAA}\;\overbrace{UUU}$$

   specifies the following amino acid sequence:

   met—pro—gly—lys—phe

   (2) There are a total of 61 codons that specify amino acids, including the "start" codon AUG, which codes for methionine. There are three "stop" codons (UAA, UAG, UGA) that signal the protein synthesis machinery to stop translation. The stop codons do not code for any amino acid.

   b. **Properties of the genetic code**

TABLE 4-4. RNA Summary

Type of RNA	Function
mRNA	Carries information for amino acid sequence from the DNA to the ribosomes
tRNA	Carries amino acids to the mRNA on the ribosome (interpreter in protein synthesis)
rRNA	Structural component of ribosomes (also some enzymatic activity)

TABLE 4-5. An Example of the Genetic Code

DNA codon	TAC	CAG	TTC	ATG
mRNA codon	AUG	GUC	AAG	UAG
Anticodon	UAC	CAG	UUC	None
Amino acid	Methionine	Valine	Lysine	Stop

- (1) **The code is redundant.** There are more than 60 different codons, but only 20 different amino acids. Two or more codons may specify the same amino acid. These redundant codons differ in the second or third position. For example, CUU, CUC, CUA, and CUG all specify the amino acid leucine.
- (2) **The code is unambiguous.** Each codon specifies one and only one amino acid.
- (3) **The code is nearly universal.** Almost all organisms share the same genetic code.

2. **Ribosomal RNA (rRNA)** does not code for a polypeptide product. The rRNA functions as a structural component of the ribosomes, the structures on which proteins are synthesized (see III).

3. **Transfer RNA (tRNA)** also does not code for protein. The tRNA functions in protein synthesis as an "interpreter" between the mRNA and amino acids. A different tRNA matches each mRNA codon with the correct amino acid.

   a. **Structure.** tRNA molecules have a distinctive cloverleaf-like structure (Figure 4-4).

   b. **Types.** There are many types of tRNA, each with a different nucleotide triplet, called the **anticodon,** in a specific region of the molecule.

   - (1) The anticodon of each tRNA can recognize a specific codon in the mRNA by complementary base pairing.

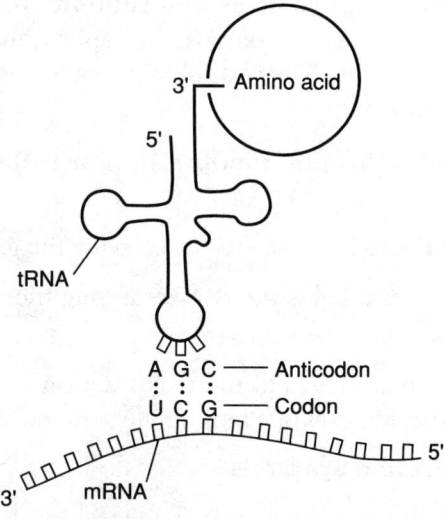

Figure 4-4. The binding of a transfer RNA (tRNA)–amino acid complex to a messenger RNA (mRNA) strand. Note the anticodon of the tRNA binds to the codon of the mRNA.

TABLE 4-6. Comparison of Transcription and Translation

	Transcription	Translation
Template	DNA	RNA
Location	Nucleus (cytoplasm in prokaryotes)	Cytoplasm; ribosomes free or on endoplasmic reticulum
Molecules involved	RNA nucleotides, DNA, RNA polymerase, transcription factors	Amino acids, tRNA, mRNA, ribosomes, enzymes, ATP, GTP, initiation/elongation factors
Enzymes needed	RNA polymerase, RNA processing enzymes	Aminoacyl-tRNA synthetase, peptidyl transferase
Control: start and stop	Transcription factors at promoter region, (TATA box), terminator region	Initiation factors, AUG, stop codons, release factors
Product	mRNA	Protein
Processing involved	RNA processing: 5' cap and poly A tail, RNA splicing (introns removed, exons joined)	Spontaneous folding, disulfide bridges, some polypeptide trimming and modifications

GTP = guanosine triphosphate.

    (2) Some tRNAs can bind to two or three redundant codons because of inaccurate pairing or "wobble" in the third nucleotide position of the tRNA.

  c. **Aminoacyl-tRNA synthetase,** an enzyme that **activates amino acids,** attaches a specific amino acid to the 3' end of each tRNA. There is a specific aminoacyl-tRNA synthetase for each amino acid, enabling it to participate in protein synthesis.

III. Translation (Table 4-6)

A. The Structure and Function of Ribosomes

1. **Structure**
   a. **Ribosomes are present as two subunits (large and small) in the cytoplasm.** Ribosomes are composed of approximately 60% rRNA and approximately 40% protein. The two subunits come together and bind the mRNA as it exits the nucleus.

   b. Each ribosome has **one binding site for mRNA** and **two binding sites for tRNA.**

   (1) The **P site** holds the tRNA carrying the growing polypeptide chain.

   (2) The **A site** holds the tRNA carrying the next amino acid to be added to the chain.

2. **Function.** The ribosomes hold the tRNA and mRNA molecules together, and enzymes catalyze the addition of a new amino acid to a growing polypeptide chain.

B. The Mechanism of Protein Synthesis

1. **Initiation** (Figure 4-5A). With the help of proteins called initiation factors, the ribosome subunits come together and bind the mRNA at the "start" codon, AUG. The

initiator tRNA (carrying methionine) binds to the start codon in the P site on the ribosome.

2. **Elongation** occurs by addition of amino acids one by one to the initial amino acid. Elongation is a three-step process that requires an input of energy from GTP hydrolysis.

   a. The anticodon of a new activated tRNA binds to the complementary codon in the A site of the ribosome (see Figure 4-5B).

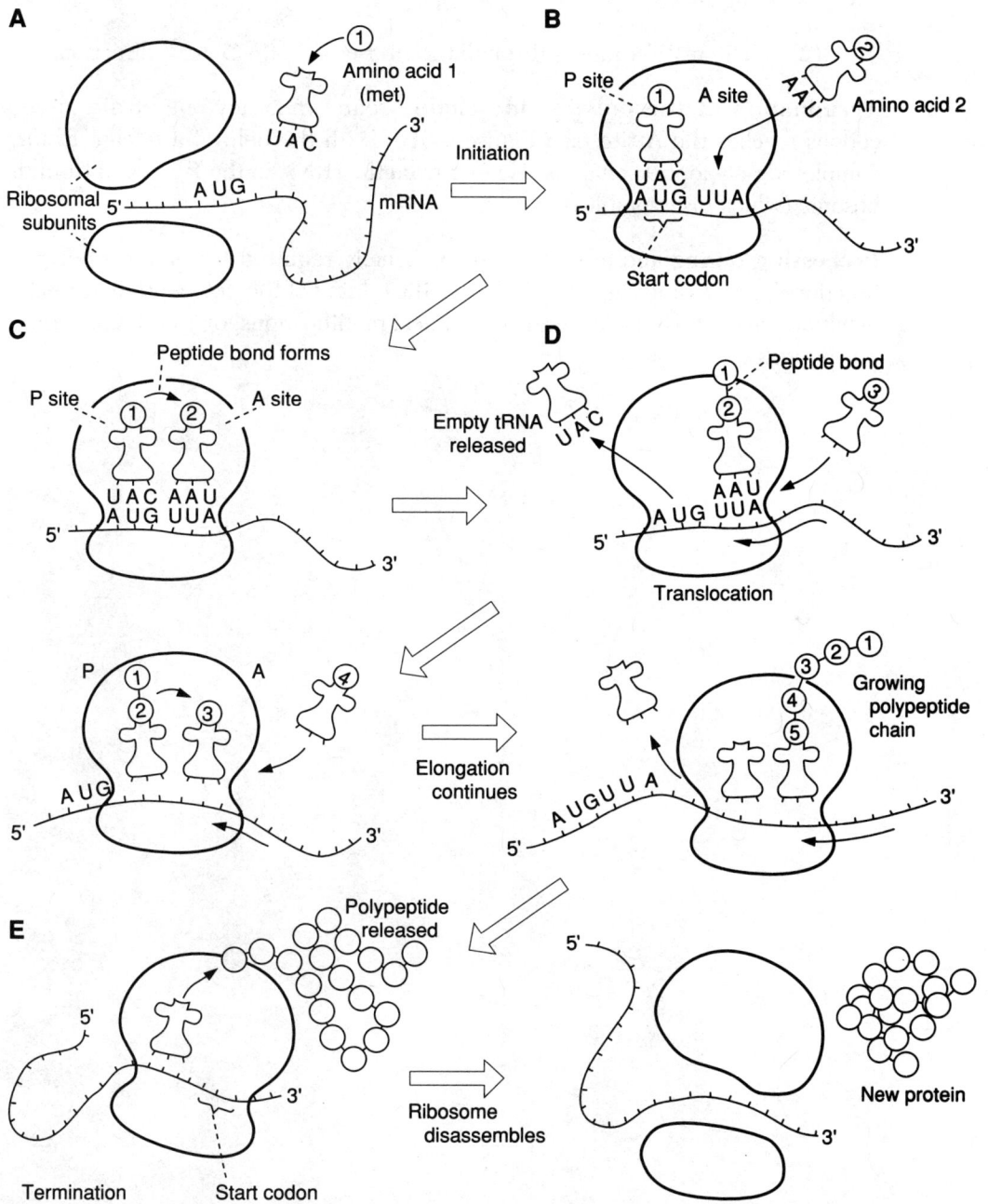

Figure 4-5. The mechanism of protein synthesis (translation). **A)** Initiation. **B) through D)** Elongation. **E)** Termination.

    b. The first amino acid in the P site (met) forms a peptide bond with the new amino acid, and the bonded amino acids are then transferred to the tRNA in the A site (see Figure 4-5C). In subsequent elongation steps, the growing polypeptide chain is transferred to the A site tRNA as each new amino acid is added.

    c. The empty tRNA is released from the P site. Then the tRNA, along with the bound mRNA, moves from the A site into the P site. This is the **translocation step** (see Figure 4-5D).

        (1) The next codon is brought into the A site, available for the correct incoming tRNA.

        (2) The mRNA moves through the ribosome in the 5′ to 3′ direction.

3. Termination of the polypeptide chain occurs when any one of the three stop codons reaches the A site (see Figure 4-5E). With the help of a release factor, the completed polypeptide chain is released from the tRNA in the P site and from the ribosome (which disassembles).

4. Processing during and after synthesis is usually required to render a polypeptide functional. Processing events include specific folding of the polypeptide, formation of disulfide bonds (tertiary structure), chemical modifications, or combining with other polypeptides.

Microbiology 5

I. Viruses

A. Viral Structure

1. Viruses are particles of genetic material surrounded by protein. A complete viral particle is called a **virion**.

    a. **Viral genomes** differ for different viruses. The genetic material may be DNA or RNA, double-stranded or single-stranded, linear or circular. Genome sizes range from just a few genes to hundreds of genes.

    b. The protein **capsid** is the outer covering of the virion.

       (1) Capsids come in various shapes, including helical, polyhedral, or complex. Some simple capsids are composed of repetitive subunits of only a single type of protein.

       (2) In some viruses, the capsid is surrounded by a lipid-membrane **envelope**. The envelope is usually derived from a host cell.

    c. Size. Viruses range from 20–300 nm in diameter. The smallest is approximately 500 times smaller than a human red blood cell; the largest is approximately the same size as the smallest bacterial cells.

B. Viral Life Histories

1. **Viruses are obligate intracellular parasites;** that is, they can reproduce only inside a living cell (the host).

    a. **Different types of viruses infect different types of cells,** including bacterial, animal, or plant cells. Once they enter a cell, viruses can command the cellular machinery to produce new virions (reproduce).

    b. **Some viruses can integrate into the genome of the host** instead of reproducing immediately after infection. The integrated viral genome can then replicate with the host's DNA. During this time, the viral infection may remain undetected.

    c. After viral reproduction, the new viruses leave the host cell to infect more cells.

2. **Bacteriophages** are viruses that infect bacterial cells. The most well studied are the T-even phages and phage lambda, which infect *Escherichia coli*.

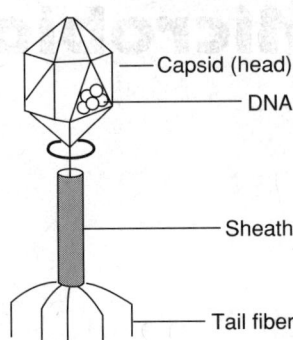

Figure 5-1. The structure of a bacteriophage.

- a. **Structure.** Bacteriophages have a complex structure composed of many types of proteins (Figure 5-1). They have an icosahedral head that encloses the genetic material (DNA), a sheath, and tail fibers.

- b. **Mechanism of infection.** A bacteriophage infects a cell by binding with its tail fibers to specific cell membrane receptors. The sheath contracts and the DNA is injected into the cell, leaving the protein component outside.

3. The **lytic cycle** is one of two alternative life cycles for bacteriophages (Figure 5-2A). **Phage lambda** illustrates the lytic cycle.

    - a. Once the viral DNA is inside the cell, it is transcribed and translated by the host cell machinery. A viral enzyme degrades the host's DNA.

    - b. When the viral DNA has been replicated and proteins have been synthesized, the new virions are assembled.

    - c. The host cell lyses and releases the phage. This entire cycle can occur in only 20–30 minutes, and it can result in a 100-fold increase in the number of phages.

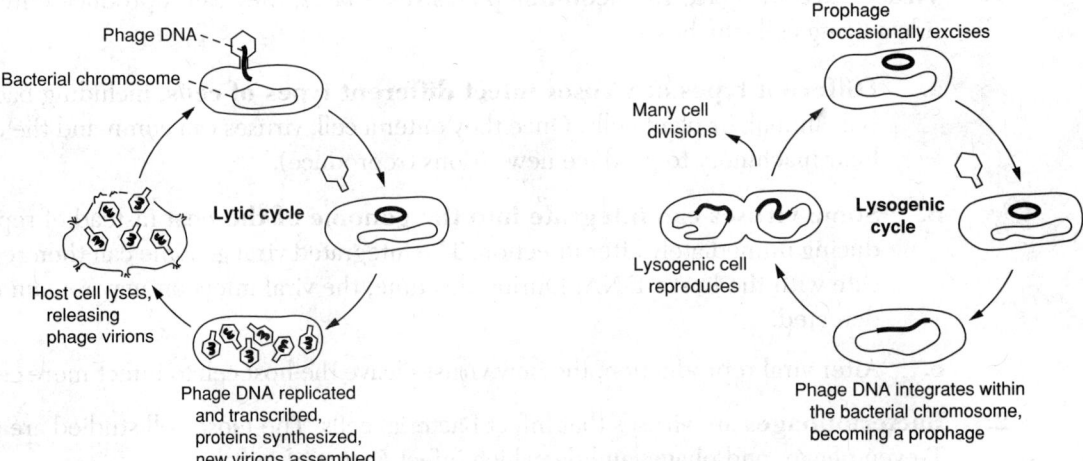

Figure 5-2. **A)** The lytic and **B)** lysogenic cycles for bacteriophages.

4. The **lysogenic cycle** is the life cycle of temperate bacteriophages (see Figure 5-2B). These phages can reproduce without killing the host cell.

   a. Once inside, the viral DNA is integrated into the host's DNA and becomes a **prophage.** During this stage, the viral genes are kept inactive by a viral repressor protein.

   b. The prophage replicates with the host chromosome and is transmitted to progeny cells during cell division. A prophage may be carried in the chromosome of the host cell for many generations. During this time, the host cell may express new properties because of the presence of the prophage.

   c. Excision of the prophage and induction of the lytic cycle may occur spontaneously or may be caused by environmental stress.

5. **Animal viruses** are viruses that infect animal cells (including human cells).

   a. **Structure.** Many animal viruses have an envelope with protruding protein spikes that bind to cell receptors. Once bound, the viral envelope fuses with the cell membrane and the entire virus (capsid and genome) enters the cell. Then, the protein capsid is removed.

   b. In the **productive cycle,** the virus reproduces soon after infection. If the genome is RNA, viral enzymes are used to replicate it and transcribe mRNA. After the host's machinery synthesizes new capsids, the new virions are assembled. They exit the host cell by budding off from its membrane, which may occur without lysing the cell.

   c. Some animal viruses with DNA genomes can integrate into the host's genome as a **provirus.** After a latent period, the virus may enter a productive cycle.

II. Prokaryotic Cells

A. **General Characteristics of Prokaryotes** (Table 5-1)

   1. **Classification.** Prokaryotes make up the kingdom *Prokaryotae* or *Monera*. All prokaryotes are bacteria (and all bacteria are prokaryotes). Bacteria are simple, unicellular organisms, and they are the most abundant organisms on Earth. All organisms other than bacteria are eukaryotes (see Chapter 6 of the Biology Review Notes for a description of eukaryotic cells).

TABLE 5-1. Summary of Prokaryotic Cell Characteristics

Characteristic	Description
Cell shape and size	Spheres, rods, or spirals; ≈1–5 μm (1/10th the size of eukaryotic cells)
Cell surface	Cell wall of peptidoglycan; may have capsule for adherence or protection; may have pili for adherence or conjugation
Motility	Move with flagella; show taxis to stimuli
Genome	Single circular DNA molecule with little associated protein; may have plasmids (1/1000th the DNA of eukaryotes)
Growth and reproduction	Divide by binary fission; some genetic exchange by conjugation

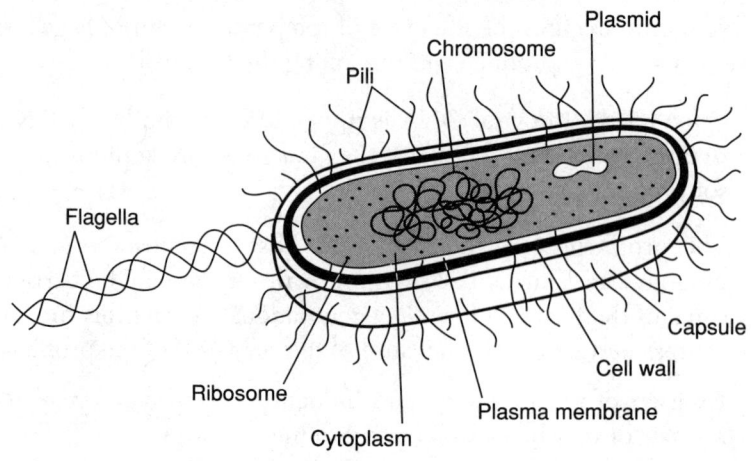

Figure 5-3. The structure of a prokaryotic cell.

2. **Structure.** Prokaryotic cells have no **membrane-enclosed nucleus.** Thus, the DNA is not in a separate compartment as it is in eukaryotic cells. This is the most fundamental structural difference between the two general types of cells.

3. The **diameter** of most prokaryotes ranges from approximately 1–10 μm, although some types can be as small as 0.1 μm. In comparison, the diameter of eukaryotic cells ranges from approximately 10–100 μm.

4. Each type of bacteria has a distinctive **cell shape.** The various shapes include rods (bacilli), spheres (cocci), and spirals (spirilla). Some bacterial cells arrange themselves in chains, clusters, or filaments of cells.

B. **Prokaryotic Cell Structure** (Figure 5-3)

1. The **prokaryotic genome** contains less DNA (approximately 1000-fold less) than the eukaryotic genome.

    a. The **bacterial chromosome** is a single, circular DNA molecule. There is very little protein associated with the DNA.

    b. **Plasmids.** Many bacterial cells also contain small, circular DNA molecules that replicate separately from the chromosome. These plasmids carry a few nonessential genes and can be transferred between bacterial cells during mating (conjugation).

2. A semipermeable **plasma membrane** surrounds the intracellular gel-like substance (cytoplasm). The plasma membrane mediates selective exchange between the cell and its external environment.

    a. Infoldings of the plasma membrane provide extra surface area for chemical reactions.

    b. There are no separate intracellular membranes or membrane-enclosed organelles in prokaryotes.

3. Most bacteria have a rigid, nonliving **cell wall** outside the plasma membrane. Bacterial cell walls usually contain a chemical called peptidoglycan, which is not found in

eukaryotes. The external wall maintains cell shape and protects the cell from the effects of osmotic changes in the environment. Some antibiotics work by damaging bacterial cell walls.

4. Some bacteria have a flexible outer covering called a **capsule,** which is exterior to the cell wall. The capsule protects cells from attackers, prevents dehydration, and allows attachment to surfaces.

5. **Pili** are short thin extensions from bacterial cells.

   a. **Common pili,** or fimbriae, are used for attachment.

   b. **Sex pili** are used for genetic transfer during conjugation.

6. **Flagella** are long structures used for motility that are attached to bacterial cells. One or many flagella may be located at one or both poles or all around the cell. Rotation of each flagella propels the cell through its fluid environment.

   a. **Taxis** is the movement response by which bacteria move toward or away from stimuli in their environment. An example is chemotaxis, which is movement toward or away from chemicals.

   b. A change in the direction of flagellar rotation (clockwise or counterclockwise) causes the cell to change its direction of movement.

7. **Endospores** are tough, metabolically inactive structures formed by some bacteria. These resting bodies are resistant to environmental stress and can survive for long periods of time. At some point, endospores can germinate to reform the vegetative (dividing) cell.

C. Prokaryotic Life Histories

1. **Binary fission** is a type of cell division by which single bacterial cells generally reproduce asexually. In this process, the single replicated chromosome is partitioned into each of the two daughter cells, which are separated by pinching in of the plasma membrane. A single cell can rapidly form an entire population of identical cells by binary fission. Most of the variation in a bacterial population results from mutations in the DNA.

2. **Conjugation** is a type of mating that bacteria undergo in which DNA is transferred between two cells that are temporarily joined (Figure 5-4). In conjugation, one cell is the donor, and the other is the recipient of genetic information.

   a. **The donor is an F+ cell** (i.e., a cell that contains a plasmid called the fertility, or F, factor). The F factor enables the F+ cell to form sex pili. **The recipient is an F− cell** (i.e., the cell lacks the F factor). The F+ cell attaches to the F− cell by its sex pili.

   b. A copy of the F factor is transferred from the donor to the recipient cell through the **cytoplasmic bridge** in the pilus. This converts the F− cell to F+. (The F+ cell remains F+.)

   c. **The F factor is an episome,** which is a plasmid that can integrate into the bacterial chromosome. When this happens, the cell becomes a **high-frequency**

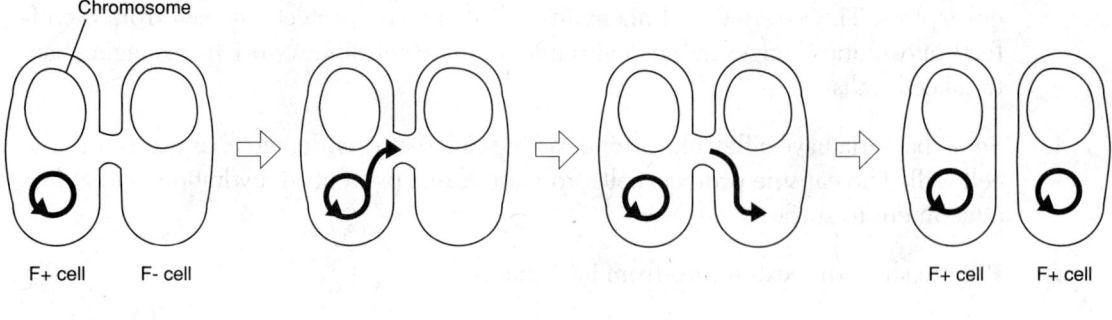

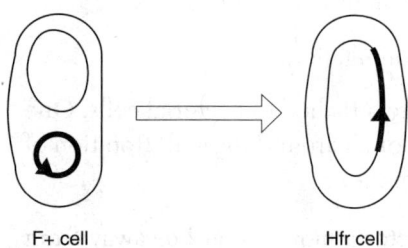

Figure 5-4. The process of conjugation in bacteria, in which DNA is transferred between two cells that are temporarily joined. F = fertility factor (+ = present; − = absent); Hfr = high-frequency recombination.

recombination (Hfr) cell. Chromosomal genes can then be transferred with the integrated F factor during conjugation.

  d. By timing the transfer of chromosomal genes during Hfr conjugation, scientists have mapped the relative locations of different genes on the bacterial chromosome.

 3. There are **two other processes that involve the transfer of DNA** in bacteria.

  a. **Transformation** occurs when bacterial cells take up DNA directly from their surroundings. Bacterial transformation is commonly used for research and biotechnology purposes.

  b. **Transduction** occurs when bacteriophages incorporate bacterial DNA into their genomes, then carry that DNA to the next host cell. There are two types of transduction.

   (1) In **general transduction,** random pieces of host DNA are picked up during the lytic cycle.

   (2) In **specialized transduction** (by temperate phages only), excision of a prophage from the host chromosome brings with it some adjacent bacterial genes.

III. Fungi

A. Characteristics of Fungi (Table 5-2)

 1. Fungi are eukaryotes; most are multicellular.

 2. All fungi obtain their organic nutrients by **absorption.** Fungi get their nutrients from various sources.

TABLE 5-2. Summary of Fungal Characteristics

Group	Examples	Morphology	Asexual Reproduction	Sexual Reproduction	Miscellaneous Information
Zygomycota	Black bread mold	Aseptate hyphae	Sporangia on aerial hyphae	Zygosporangia	In soil and decaying matter
Ascomycota	Sac fungi, yeasts	Septate hyphae	Conidia	Ascosporangia, asci	Plant parasites, lichens, decomposers
Basidiomycota	Club fungi, mushrooms, shelf fungi	Dikaryotic hyphae	Uncommon	Basidia	Some edible, plant parasites
Deuteromycota	Penicillium	Molds	Conidia	None known	Some produce antibiotics

    a. Most fungi are **parasitic**—they absorb nutrients from the body of a living host.

    b. Some fungi are **saprophytic**—they absorb nutrients from dead organic matter (i.e., the fungi are decomposers).

    c. Some fungi are **mutualistic**—they absorb nutrients from a host but reciprocate to benefit the host.

3. There are three major **types of fungi** that are distinguished by their structure.

    a. **Molds** are multicellular, filamentous organisms, such as mildew, rust, and smut.

    b. **Fleshy fungi** are multicellular, filamentous organisms that produce a thick (fleshy) reproductive body. The fleshy fungi include mushrooms, puffballs, and coral fungi.

    c. **Yeasts** are nonfilamentous, unicellular organisms, typically spherical or oval in shape.

B. Fungal Structures

1. **Vegetative structures** are composed of cells involved in catabolism and growth.

    a. **Hyphae** are long filaments of cells joined together to form the body (thallus) of a mold or fleshy fungus. A hyphae grows at its tips to form an intertwined mass of hyphae called a **mycelium.** Each part of the vegetative mycelium is capable of growth.

    b. **Septa** are the crosswalls that divide the hyphae of most fungi into uninucleate cells. However, the hyphae of a few fungi do not contain septa (i.e., they are aseptate). These **coenocytic hyphae** are composed of long, continuous, multinucleate cells.

    c. Most nuclei of fungal mycelia are **haploid.**

    d. Fungi have no flagellated structures and are **nonmotile.**

2. The **reproductive or aerial mycelium** is the part of the fungus concerned with reproduction. The aerial mycelium produces **spores,** the reproductive structures in both asexual and sexual reproduction.

    a. Spores are generally produced asexually in favorable conditions and asexually in unfavorable conditions (e.g., stress).

    b. Spores are dispersed by wind or water. They germinate in favorable conditions to reform the vegetative body.

C. General Life Histories of Fungi

1. **Asexual spores** are most frequently produced by the aerial mycelium of one haploid organism. The spores are produced through mitosis and subsequent cell division. When these spores germinate, they become organisms that are genetically identical to the parent. Note that each of the three divisions of fungi form their haploid, asexual spores in different ways.

    a. The *Zygomycota* division produces asexual spores in sacs called **sporangia** at the tips of aerial hyphae.

    b. Most fungi of the *Ascomycota* division produce chains of spores called **conidia**.

    c. The *Basidiomycota* division does not undergo asexual reproduction.

2. **Budding** is a type of asexual reproduction in which the parent cell forms a protuberance (bud) that elongates and eventually breaks away as an independent daughter cell. **Yeasts** belong to the *Ascomycota* fungi group, but differ in their reproduction. Yeasts are colonies of unicellular organisms that do not form asexual spores but reproduce asexually by budding.

3. **Fungal sexual spores** are formed by sexual reproduction.

    a. The following steps are involved.

       (1) Two opposite mating strains of the same species of fungus conjugate by fusion of hyphae from each of the organisms.

       (2) A haploid nucleus of the donor cell enters the recipient cell, and the two nuclei fuse. This step is delayed in some fungi, which results in cells with two nuclei (dikaryons). Fusion of the nuclei results in a diploid sexual spore.

       (3) The sexual spore divides by meiosis to form new haploid spores. The spores can then form new haploid mycelia.

    b. **Methods of sexual spore production.** The various fungi produce sexual spores in different ways.

       (1) *Zygomycota,* such as black bread mold, form resistant bodies (**zygosporangia**) that can remain dormant in unfavorable environmental conditions.

       (2) *Ascomycota* (sac fungi) produce sexual spores in a saclike structure called an **ascus.**

       (3) *Basidiomycota* (club fungi), such as mushrooms, produce sexual spores externally on a base pedestal called a **basidium.**

4. ***Deuteromycota,*** or **imperfect fungi,** is a group of fungi with no known sexual reproduction. Their asexual reproduction is by conidia. An example is *Penicillium*.

5. **Lichens** are symbiotically associated fungi and algae. They have a mutualistic relationship in which each benefits the other.

The Eukaryotic Cell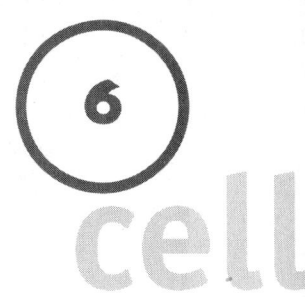

I. Overview

A. All Organisms Other than Bacteria Have Eukaryotic Cells

Eukaryotic organisms may be unicellular or multicellular. Eukaryotic cells range in size from approximately 10–100 μm in diameter (approximately 10 times larger than prokaryotic cells).

B. Eukaryotes Are More Advanced in an Evolutionary Sense than Prokaryotes (See Chapter 20.) The fundamental difference between the two cell types is that **eukaryotic cells have a membrane-bound (true) nucleus.**

1. The nucleus is the information center of the cell. It contains the genes, which are encoded in DNA, that control all of the cell activities.

2. The **cytoplasm** of the cell includes the entire region outside the nucleus. Cytoplasmic structures are suspended in a gel-like fluid called **cytosol**.

3. **Membranes** are important components of most eukaryotic cell structures.

    a. **Organelles** are intracellular structures surrounded by one or more membranes. The normal function of a cell involves several types of membrane-bound organelles.

    b. **Intracellular membranes** provide internal surface area and partition the cell into separate compartments. Specific chemical reactions occur at the membrane surfaces and within the compartments.

C. Cell Structure Can Only Be Studied by Microscopic Observation

1. **Light microscopes** are used for magnification of living cells up to 1500-fold.

2. **Electron microscopes** are used to magnify specially prepared (dead) cells up to 250,000-fold.

II. The Eukaryotic Cell: Structures and Functions (Figure 6-1; Table 6-1)

A. Nucleus

1. The nucleus is a membrane-bound organelle containing most of the genetic material of the cell.

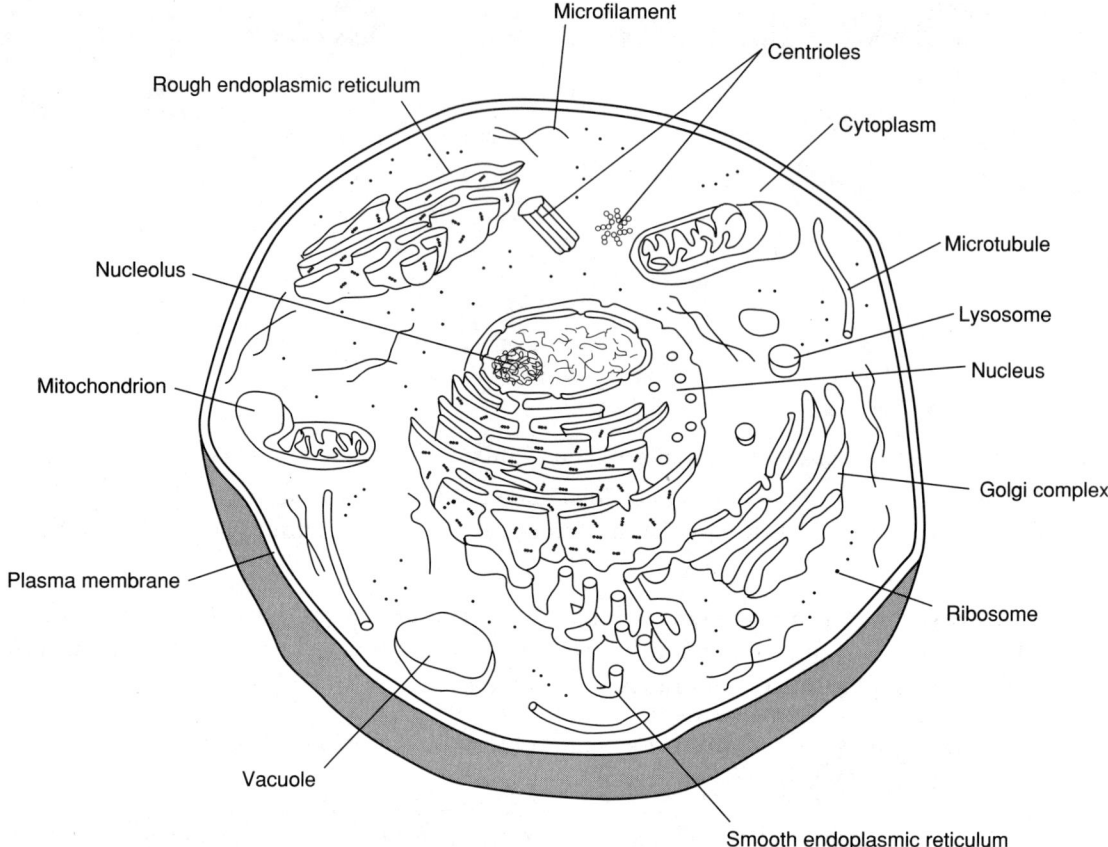

Figure 6-1. The structure of a eukaryotic cell.

2. **Genetic material** resides in the nucleus in the form of **chromatin,** which is a complex of DNA and associated histone proteins. Chromatin appears as wispy, indistinct material in the nucleus of nondividing cells.

    a. **Chromosomes** become visible when a cell prepares to divide and the chromatin condenses. Chromosomes are the distinct, linear structures that carry the genes. Each species of organism has a characteristic number of chromosomes in the nucleus (e.g., humans have 46).

    b. During **mitosis,** the two copies of each chromosome are separated into two sets, which are segregated into the daughter cells.

3. The **nucleolus** is the most conspicuous structure in a nondividing cell. This is a special region of DNA where the genes for ribosomal RNA (rRNA) are located. The nucleolus is also where the rRNA and ribosomal proteins assemble to form the ribosomes.

4. The **nuclear envelope** surrounds the nucleus with a double membrane. Molecules are exchanged between the nucleus and the cytoplasm through **pores** in the nuclear envelope. A complex of proteins is organized around each pore.

B. Membrane-Bound Organelles

1. The **mitochondria** serve as the sites of cellular respiration, the oxygen-requiring catabolic process that extracts energy from organic molecules to produce adenosine

Table 6-1. Summary of Eukaryotic Cell Structures

Structure	Description	Functions
Membrane-bound nucleus	Large orangelle surrounded by double membrane with pores; contains chromosomes (DNA and histones)	Information storage and transfer
DNA	Multiple, linear chromosomes associated with histone proteins; usually in dispersed form (chromatin)	Heritable material
Mitochondria	Double-membrane bound with highly folded inner membrane, ETC	Cellular respiration (ATP production)
Lysosomes	Membrane sacs with hydrolytic enzymes	Digestion, organelle recycling
Vacuoles/Vesicles	Membrane sacs (various sizes and compositions)	Transport, chemical reactions, water balance, food intake
Endoplasmic reticulum (rough and smooth)	Membrane sacs and tubules	Manufacture of export proteins, membrane proteins and lipids
Ribosomes	Two protein/RNA subunits; free in cytoplasm or attached to ER membrane	Protein synthesis
Golgi apparatus	Stacked, flattened membrane sacs	Product modification, packaging & transport
Plasma Membrane	Phospholipid bilayer with proteins, also cholesterol and carbohydrates	Selective exchange with environment
Cytoskeleton	Complex network of three filament types	Structural support, shape, movement
Cilia and flagella	Cytoplasmic extensions with a "9+2" arrangement of microtubules	Cell motility
Centrioles	Self-replicating microtubule structures	Cell division

*Structure is not found in all cell types.
ATP = adenosine triphosphate; DNA = deoxyribonucleic acid; ETC = electron transport chain.

triphosphate (ATP). The mitochondria are sometimes referred to as the "powerhouse of the cell." Mitochondria are found in nearly all eukaryotic cells, in varying numbers depending on the metabolic activity of the cell. They are dynamic organelles, capable of growth and division.

    a. **Membranes.** Mitochondria are surrounded by two membranes—one inner and one outer.

        (1) The **inner membrane** is highly convoluted, forming many infoldings called **cristae.** The cristae provide surface area for the chemical reactions of respiration. Electron transport occurs here.

        (2) The **outer membrane** is smooth and is in contact with the cellular cytoplasm.

        (3) The **intermembrane space** exists between the inner and outer membranes. This space is important in separating and insulating inner mitochondrial structures.

        (4) The **matrix** is the compartment that is internal to the inner membrane.

This compartment contains many metabolic enzymes. The Krebs cycle occurs here.

    b. **Enzymes.** Specific mitochondrial enzymes and other proteins are localized in each compartment of the cell and are embedded within the membranes. Mitochondria contain some DNA and ribosomes to program and carry out the synthesis of some of their own proteins.

2. **Lysosomes** are membrane-bound organelles that contain hydrolytic (digestive) enzymes. The inside of the lysosome is highly acidic, which is the optimal environment for hydrolytic enzymes. Lysosomes serve several important functions in cells.

    a. Lysosomes are needed for **digestion** of food or other substances that are engulfed by the cell. These substances are pinched off into a membrane-bound **food vacuole.** When a lysosome fuses with the vacuole, the lysosomal enzymes digest the food.

    b. **Recycling of organic materials** involves lysosomal destruction of old or damaged organelles. In "storage diseases," such as Tay-Sachs disease, defects in lysosome function result in the accumulation of substances that interfere with cell function.

    c. **Programmed cell destruction** occurs during embryonic development to give body parts their proper form (e.g., separated fingers).

3. **Vacuoles and vesicles** include various types of membranous sacs, which differ in size (vacuoles are larger than vesicles) and function.

    a. **Transport vesicles** carry enclosed substances from place to place in the cell.

    b. **Food vacuoles** contain solid materials that are brought into the cell from its surroundings.

    c. **Water vacuoles,** also called contractile vacuoles, are found in many freshwater protozoa. These organelles pump out excess water so that the cell does not burst.

    d. **Central vacuoles** are found in plant cells, where they function in storage, digestion, pigment localization, and water absorption. They are the most conspicuous organelle in a plant cell.

C. Intracellular Membranes

1. **Membranous structures** are a complex, constantly changing network found within the cytoplasm of the eukaryotic cell. These structures interact either directly through physical contact or indirectly through the exchange of vesicles. Each type of membranous structure provides a distinct local environment for specific chemical reactions.

2. The **endoplasmic reticulum (ER)** is an organization of membranous sacs and tubules that is connected to the nuclear envelope. The ER compartments or **cisternae** sequester an internal lumen (the cisternal space) from the cytosol. There are two types of ER—rough and smooth.

    a. The **rough ER** is made up of sacs and folds of membrane, which is studded with attached ribosomes. The rough ER is involved in the manufacture and transport of proteins that are secreted from the cell (export proteins) or used in cell membranes.

(1) A **signal sequence** is a specific sequence that starts the messenger RNAs, which code for export proteins. The signal sequence makes it possible for the ribosomes translating these RNAs to link to specific receptors on the rough ER. All other ribosomes remain free in the cytoplasm. As the export protein is being synthesized, it enters the cisternal space, and the signal sequence is enzymatically removed.

(2) **Glycoproteins** used in the plasma membrane are also produced in the ER by enzymatic attachment of a sugar to specific proteins.

(3) ER products eventually depart in **transport vesicles** that are pinched off from the membrane. These vesicles carry the materials out of the cell, to the plasma membrane, or to intracellular destinations.

b. The **smooth ER** is made up of tubules of naked membrane. The reactions in the smooth ER are involved in **lipid metabolism.**

(1) **Phospholipids, fats, and steroids are synthesized** in the smooth ER.

(2) The smooth ER is also where certain drugs, such as alcohol and barbiturates, are **detoxified.** Chemical modification by enzymes in the smooth ER allows these drugs to be flushed from the body.

3. The **Golgi apparatus** is a series of flattened and stacked membrane sacs. These structures are involved in the chemical modification, storing, and distribution of products made in the ER.

a. The *cis* face of the Golgi apparatus **receives transport vesicles from the ER.** Golgi enzymes chemically modify the received materials (e.g., by adding an oligosaccharide to some glycoproteins). Other compounds are synthesized in the Golgi apparatus itself.

b. The **chemical modifications** in the Golgi apparatus provide organic materials with "address tags" for their transport to a particular destination in the cell. The products are enclosed in vesicles that pinch off into the cytoplasm from the *trans* face of the Golgi apparatus.

D. **Production of a Secretory Product**

The following summarizes the production of a secretory product.

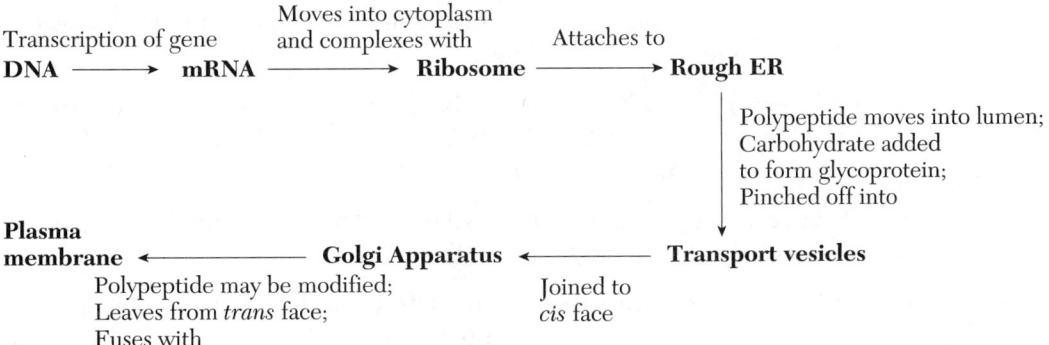

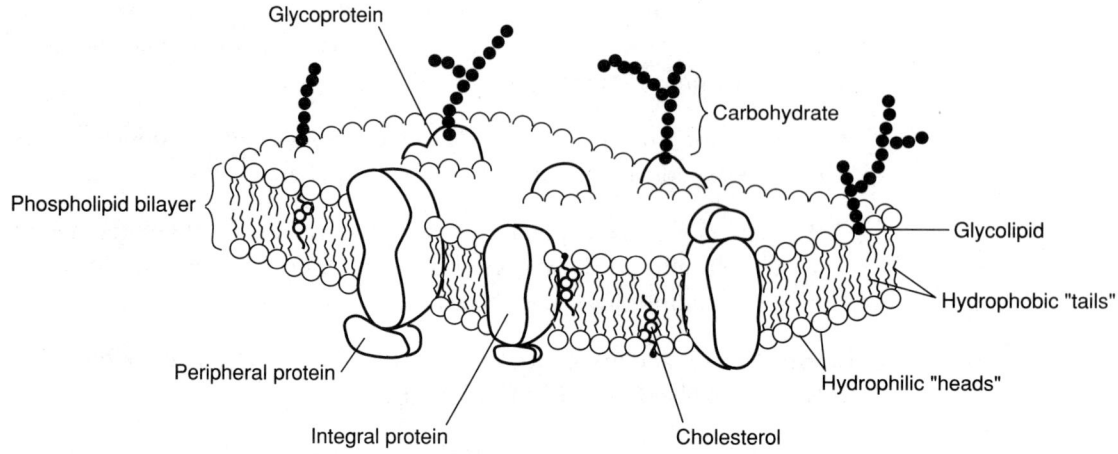

Figure 6-2. The structure of a plasma membrane.

E. Plasma Membrane

1. **Protective function.** The plasma membrane surrounds the entire cell, providing a physical barrier between the cytoplasm and the extracellular environment. The membrane is **selectively permeable,** so that only certain substances can cross it. This allows for controlled cell–cell and cell–environment communication and exchange (see II D 3).

2. **Plasma membrane structure** (Figure 6-2)

    a. **The plasma membrane is a phospholipid bilayer** (as are the intracellular membranes). The membrane is a very thin sheet composed of two layers of organic molecules called phospholipids. Each phospholipid has a polar "head" at one end (a negatively charged phosphate group) and a nonpolar "tail" at the other end (a long, hydrocarbon chain). These chemical properties lead to spontaneous formation of the bilayer structure in aqueous fluids.

    (1) **The polar head is hydrophilic** (i.e., water loving) and points toward the exterior of the bilayer. This surface comes in contact with the aqueous cytoplasm and extracellular fluids.

    (2) **The nonpolar tail is hydrophobic** (i.e., water hating) and points toward the interior of the bilayer. This region is sequestered away from the cytoplasm and extracellular fluids.

    b. The **fluid consistency** (much like salad oil) of the membrane is caused by the weak hydrophobic interactions of the bilayer.

    (1) **Phospholipid molecules can drift laterally** in the membrane and change places with their neighbors.

(2) **Cholesterol and unsaturated fatty acids** are included in the membrane bilayer. These bulky lipids prevent the plasma membrane from hardening in cold environments.

c. **Proteins** make up approximately 60% of the plasma membrane (40% is lipid). Proteins are inserted individually into the bilayer and are capable of some lateral drift. This structural design of the membrane is described by the **fluid mosaic model.**

   (1) **Types**

   (a) **Integral proteins** penetrate the interior hydrophobic zone of the membrane. Most integral proteins span the membrane, with their hydrophilic regions protruding from one or both surfaces.

   (b) **Peripheral proteins** are located on the membrane surface, anchored by attachment to integral proteins. Some peripheral proteins are on the extracellular side of the plasma membrane and some are on the intracellular side.

   (2) **Synthesis.** Plasma membrane proteins are synthesized on the rough ER, where they are inserted directly into the ER membrane. After being transported in vesicles to the cell surface, the rough ER membrane (including proteins) fuses with the plasma membrane. The proteins that face the cytoplasmic side of the ER membrane are destined to face the cytoplasmic side of the plasma membrane as well. Likewise, the proteins that face the lumen of the ER will face the extracellular side of the plasma membrane.

3. **Movement across the membrane**

   a. **Overview**

   (1) The plasma membrane is **selectively permeable,** allowing some substances to cross more easily than others.

   (2) **Nonpolar molecules** and **small polar molecules** can pass through the phospholipid bilayer via passive transport. Examples include oxygen, carbon dioxide, and water molecules.

   (3) **Large molecules and polar molecules** cannot pass through the hydrophobic zone of the phospholipid bilayer. Examples include glucose and charged molecules, such as ions. These substances must be carried across the membrane by selective transport proteins or **carrier proteins.**

   b. **Passive transport**

   (1) A **concentration gradient** is the driving force for movement of a substance by passive transport. When molecules of a solute are more concentrated on one side of the membrane than on the other, they move spontaneously from the area of higher concentration to the area of lower concentration (i.e., down the concentration gradient).

   (a) The **rate of movement** is regulated by the permeability of the membrane to the particular substance. Water diffuses freely across most cell membranes.

   (b) **Cellular energy is not needed** for passive transport.

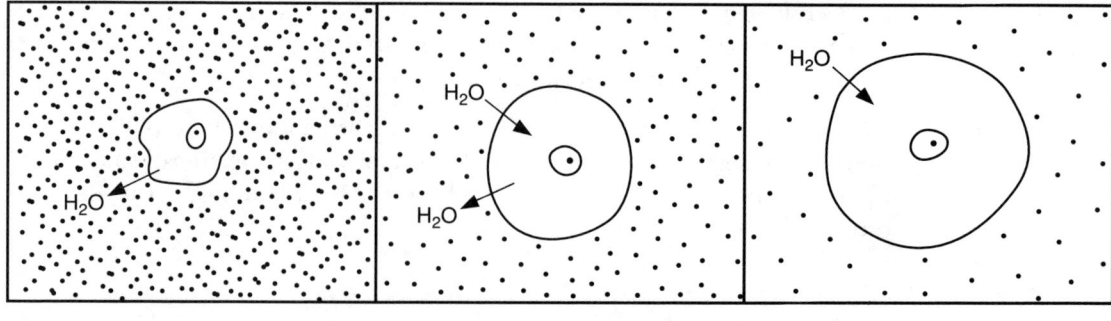

A. Hypertonic environment   **B. Isotonic environment**   **C. Hypotonic environment**

Figure 6-3. **A)** The effects on water movement and cell size of a hypertonic environment. **B)** An isotonic environment. **C)** A hypotonic environment.

- (2) **Osmosis** is a special case of passive transport. Osmosis allows the diffusion of water molecules across a membrane that is permeable to water but not to the solutes in the water. Cytoplasm and extracellular fluids are both aqueous solutions.

  - (a) **Water diffuses** from an area of lower solute concentration to an area of higher solute concentration. The net movement of water is **down its concentration gradient.**

  - (b) **Osmotic pressure** is the tendency for a solution, regardless of the type of solute, to take up water by osmosis. Osmotic pressure depends on the total solute concentrations, regardless of whether there are different solutes on each side of a membrane.

  - (c) **Tonicity** concerns the relation of a cell with its environment.

    - (i) An extracellular environment is **hypertonic** to a cell when the concentration of solutes outside the cell is greater than inside. In a hypertonic environment, the cell loses water and shrinks (Figure 6-3A).

    - (ii) An environment is **hypotonic** to a cell when the solute concentration outside is **less** than inside. In a hypotonic environment, water enters the cell, causing it to swell and even to pop (see Figure 6-3B).

    - (iii) A cell and its environment are **isotonic** when the solute concentration is the same both inside and outside the cell. In this case, water molecules move across the membrane at the same rate in both directions (see Figure 6-3C). The system is in **dynamic equilibrium.**

- (3) **Facilitated diffusion** is a type of passive transport in which solute molecules are moved down a concentration gradient by **carrier proteins** in the membrane. A carrier protein picks up a molecule, then shifts to an alternative conformation to deposit the molecule on the other side of the membrane (Figure 6-4). **No input of energy** is required.

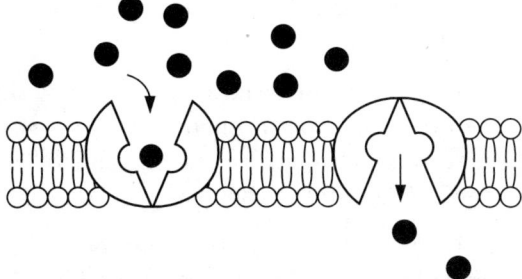

Figure 6-4. In facilitated diffusion, a passive process, a carrier protein specifically transports molecules across cell membranes.

   c. **Active transport**

   (1) Active transport moves small molecules up a concentration gradient. The driving force for movement by active transport is an **input of cellular energy.**

   (2) Active transport involves **carrier proteins** that are linked to a source of energy, such as ATP hydrolysis.

   (3) Certain carrier proteins called **ion pumps** use active transport to generate and maintain gradients of ion concentrations across a membrane.

   (a) An example of an ion pump is the **sodium–potassium pump—ATPase** (Figure 6-5). This protein (i.e., ATPase) uses ATP to pump sodium out of the cell and potassium into the cell. For every three sodium ions pumped out, two potassium ions are pumped in. This results in a **voltage gradient** caused by the net negative charge (voltage) inside the cell. The sodium–potassium pump also regulates intracellular solute concentration by keeping the intracellular sodium low. This function helps to maintain osmotic balance with the environment.

   (b) Each type of ion moves down its electrochemical gradient by passing through specific **ion channels** in the membrane. Electrostatic

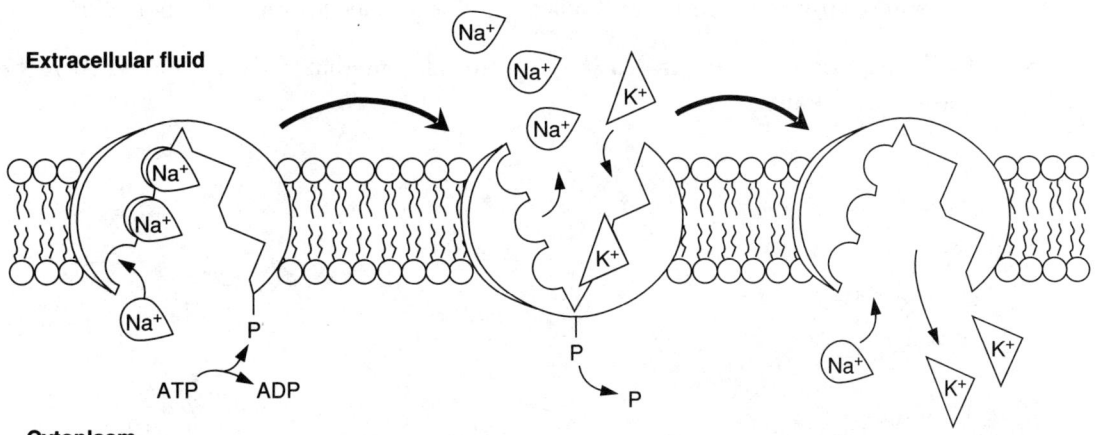

Figure 6-5. The sodium–potassium ATPase. Three sodium ions are pumped out of the cell for every two potassium ions pumped into the cell. This process requires the input of adenosine triphosphate (ATP) for energy.

attractions favor movement of cations into the cell and movement of anions out of the cell.

- (c) **Membrane potential** is generated by the differences in ion concentrations across a membrane. This is an essential property for electrically excitable cells such as nerve and muscle cells (see Chapter 7 of the Biology Review Notes).

d. **Endocytosis.** Large molecules are brought into cells by endocytosis, in which the **plasma membrane buds inward to form vesicles that pinch off into the cytoplasm.** There are three types of endocytosis.

- (1) In **phagocytosis** ("cell-eating"), an extension of the plasma membrane brings in solid particles, such as food, enclosing them in a food vacuole (Figure 6-6A). The vacuole must fuse with a lysosome for digestion to occur.

- (2) In **pinocytosis** ("cell-drinking"), droplets of extracellular fluid are incorporated into small vesicles (see Figure 6-6B). This is a nonspecific process in which all the solutes in the droplets are taken in.

- (3) **Receptor-mediated endocytosis** is the specific uptake of molecules that bind to clusters of receptors in specific regions of the membrane, called coated pits (see Figure 6-6C). The "budding in" of a coated pit forms a vesicle (reinforced by a fibrous protein called clathrin), which then releases the ingested material. This process allows cells to take in large amounts of a specific substance, even if that solute has a low extracellular concentration.

e. **Exocytosis.** Large molecules are removed from a cell by fusion of cytoplasmic vesicles with the plasma membrane. This process is used by secretory cells to export products to the extracellular environment.

4. **Receptors** are specific membrane proteins that face the outside of the cell and bind substances in the cell surroundings. Different cell types have distinct sets of membrane receptors. Each type of receptor binds and responds to a specific set of molecules (the ligands) in a characteristic way. Substances that bind receptors include hormones, viruses, bacteria, antibodies, and the surface proteins of other cells.

5. **Cell–cell recognition and adhesion** are also mediated by components of the plasma membrane.

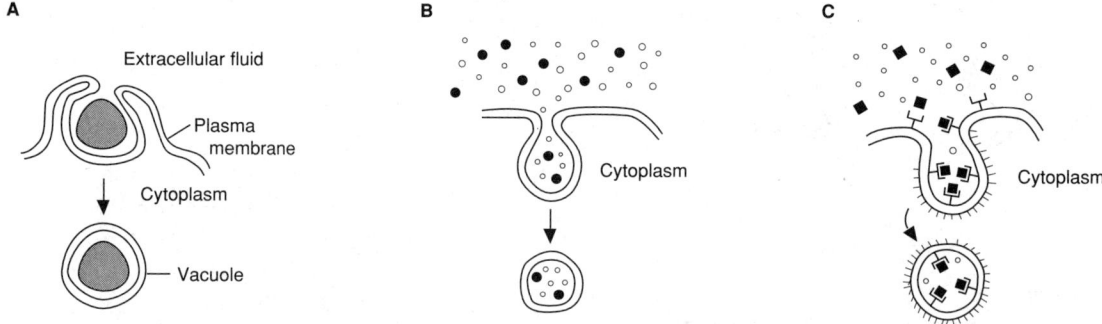

Figure 6-6. The process of **A)** phagocytosis, **B)** pinocytosis, and **C)** receptor-mediated endocytosis.

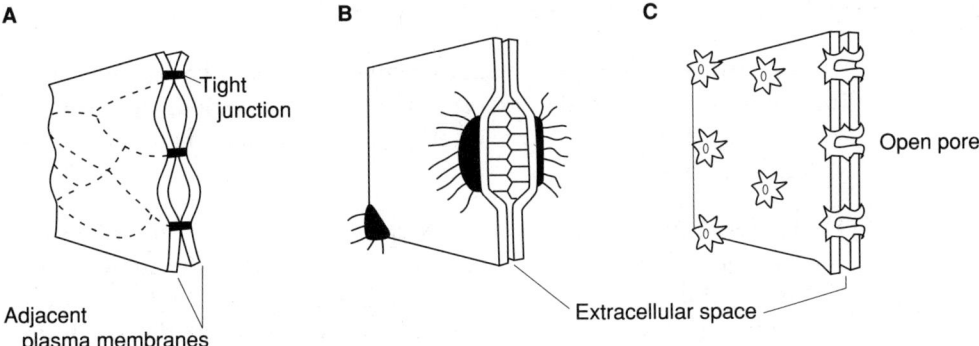

Figure 6-7. Three forms of intercellular junctions: **A)** tight junctions, **B)** desmosomes, and **C)** gap junctions.

- (a) **Glycoproteins** are proteins with attached carbohydrates, such as branched polysaccharides. Glycoproteins identify cells as belonging to a particular organism or tissue rather than to a foreign invader. For example, glycoproteins are the blood-type determinants that prevent an immune response against a person's own red blood cells.

- (b) **Intercellular junctions** formed between adjacent cells mediate direct cell–cell interactions in multicellular organisms. There are three major types.

    - (1) **Tight junctions** are connections between the adjacent plasma membranes that do not leave enough space for substances to cross the cell surfaces (Figure 6-7A).

    - (2) **Desmosomes,** or spot welds, are rigid junctions that allow transport of substances through the space between adjacent cells and across the cell surfaces (see Figure 6-7B).

    - (3) **Gap junctions** are pores between adjacent cells that allow transfer of small molecules between the two cytoplasms (see Figure 6-7C).

F. Cytoskeleton (Table 6-2)

  1. **Structure.** The cytoskeleton of the cell is composed of a dynamic network of fibers organized throughout the cytoplasm. The cytoskeleton is important for structural support, control of cell shape, cell motility, and intracellular movement.

Table 6-2. Cytoskeleton Summary

Element	Structure/Size	Protein	Functions
Microtubules	Hollow tubes largest (25 nm diameter)	Tubulin (dimers)	Motility, cell shape, chromosome & organelle movement, intracellular organization
Microfilaments	Twisted strands smallest (7 nm diameter)	Actin	Muscle cell contraction, cytoplasmic streaming, amoeboid movement, division furrow, cell shape changes
Intermediate filaments	Hollow tubes medium (8-10 nm diameter)	Varies	Structural support, tension, cell shape maintenance

2. **Types.** There are three major types of cytoskeletal elements.

   a. **Microtubules** are long hollow tubes composed of subunits of the protein **tubulin.** Microtubules have the largest diameter of the various cytoskeletal fibers. Microtubule structures are organized by regions called microtubule-organizing centers (Figure 6-8).

      (1) **Microtubules can move objects.** Microtubules can be lengthened or shortened by the spontaneous addition or removal of tubulin subunits at their ends. This process can cause objects attached to the microtubules to move. An example of this type of movement occurs during chromosome segregation in mitosis.

      (2) **Microtubules provide transportation "tracks"** in the cytoplasm. Objects move by attaching to specific proteins that can slide along these microtubule tracks. An example of this is the movement of transport vesicles from the Golgi apparatus to the plasma membrane for protein export.

      (3) **Cell motility** is another important function of microtubules.

   b. **Microfilaments** are solid fibers made of two intertwined chains of the protein **actin.** Microfilaments have the smallest diameter of the cytoskeletal fibers. Microfilaments are important for cellular movement that involves **contractions.** Examples include:

      (1) Contraction of muscle cells

      (2) Ameboid movement, involving extension and contraction of a part of a cell

      (3) Cytoplasmic streaming, involving mass movement of cytoplasm (seen in large plant cells)

      (4) Formation of the cleavage furrow during cell division

   c. **Intermediate filaments** have varying protein compositions and sizes in different cell types. **Cellular stability** is thought to be the main function of intermediate filaments. For example, intermediate filaments hold organelles in place and help maintain cell shape.

3. **Cilia and flagella.** The cilia and flagella are the microtubule structures responsible for cell motility. These structures are long, thin cell protrusions that are continuous with the plasma membrane (see Figure 6-8).

   a. **Structure.** Cilia and flagella are composed of nine sets of microtubule triplets arranged in a ring with a pair of microtubules in the center ("9 + 2 arrangement"). The base or anchor of the structure is called the **basal body,** which is composed of nine sets of microtubules, with no additional pair in the center.

      (1) Flagella are longer than cilia.

      (2) There is usually only one or a few flagella per cell, whereas cilia are much more numerous.

      (3) Eukaryotic cilia and flagella move in a wavelike motion.

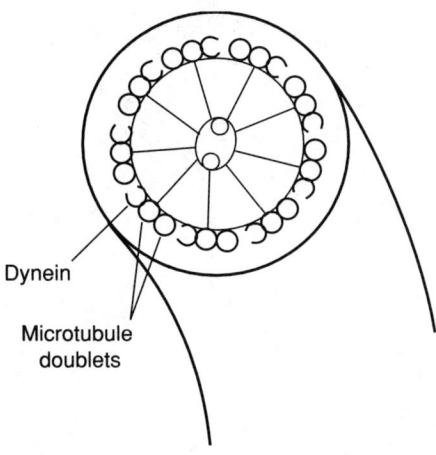

Figure 6-8. A cross section demonstrating cilia and flagella structure.

    b. Function. The most common function of the cilia or flagella is to propel a cell through its fluid environment. However, some ciliated cells in multicellular organisms are nonmotile; they use their cilia to sweep materials across the surface of a tissue.

    c. Movement. Movement of cilia or flagella involves a protein called **dynein,** which is associated with the microtubules. Dynein attaches to one microtubule and causes it to slide in a ratchet-like motion along an adjacent microtubule. This process requires an input of energy (ATP).

4. **Centrioles.** Centrioles have essentially the same microtubular structure as basal bodies, but they are not associated with cilia or flagella. A pair of centrioles is associated with each of two microtubule-organizing centers in dividing cells. Nondividing cells have only one pair of centrioles.

    a. Animal cells have centrioles, but plant cells do not.

    b. Centrioles are involved in the movement of chromosomes during mitosis (see III).

III. The Eukaryotic Cell Cycle and Mitosis

In **multicellular organisms,** cell division plays a primary role in **growth, reproduction, and repair.** Many cells divide during a period of growth, but they lose that ability as they become specialized for a particular function in the mature organism. In **unicellular organisms,** cell division is the **means for the organisms to reproduce themselves.** The **cell cycle** is the interval between formation of a cell and division of that cell to form two new daughter cells (Figure 6-9). Typical cell cycles range from 1–24 hours, depending on cell type, age, and conditions. The cell cycle includes two phases: **interphase** and **mitosis.**

A. Interphase accounts for approximately 90% of the cell cycle. Growth and metabolic activities occur continuously during interphase. The nuclear genetic material is dispersed as chromatin.

    1. During interphase, **cells grow** by increasing in size and by increasing the number of organelles in preparation for division.

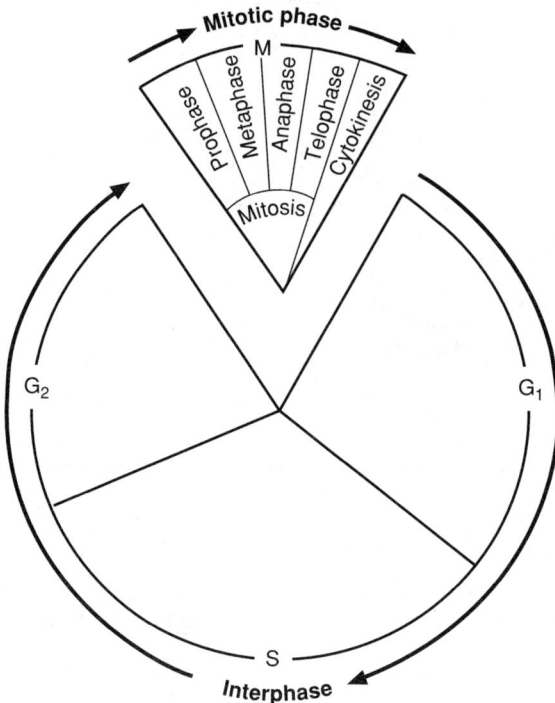

Figure 6-9. The cell cycle.

2. Interphase is composed of a **continuous succession of three phases.**

   a. During the **gap 1 (G1) phase,** metabolic activities resume at a high rate after having slowed during mitosis.

   b. **DNA synthesis** occurs in **synthesis (S) phase.** At the end of S phase, the DNA content of the cell has doubled and all the chromosomes are replicated. The single pair of centrioles has also duplicated, forming two pairs.

   c. During **gap 2 (G2) phase,** the cell makes final preparations for division.

B. **Mitosis (M) Phase** is the process by which the chromosomes are compacted then separated into two equal sets. Mitosis is usually followed by cytokinesis, which is division of the cytoplasm to form the two identical daughter cells. Mitosis is composed of a continuous succession of several stages: **prophase, metaphase, anaphase, and telophase** (Figure 6-10).

1. **Early prophase**

   a. **The DNA condenses to form discrete, visible chromosomes.** Having been duplicated during S phase, each chromosome is composed of two identical copies called **sister chromatids.** Each chromatid is one continuous DNA molecule (with associated histones). The two chromatids are joined at a constricted region of the chromosome called the **centromere.** The chromosome "arms" extend from the centromere and terminate at structures called **telomeres.**

   b. **The nucleoli** (dark regions) in the nucleus **disappear.**

   c. **The mitotic spindle,** which is a bipolar organization of the cytoskeleton, forms in the cytoplasm. The spindle starts to form at the **centrosome,** which is the

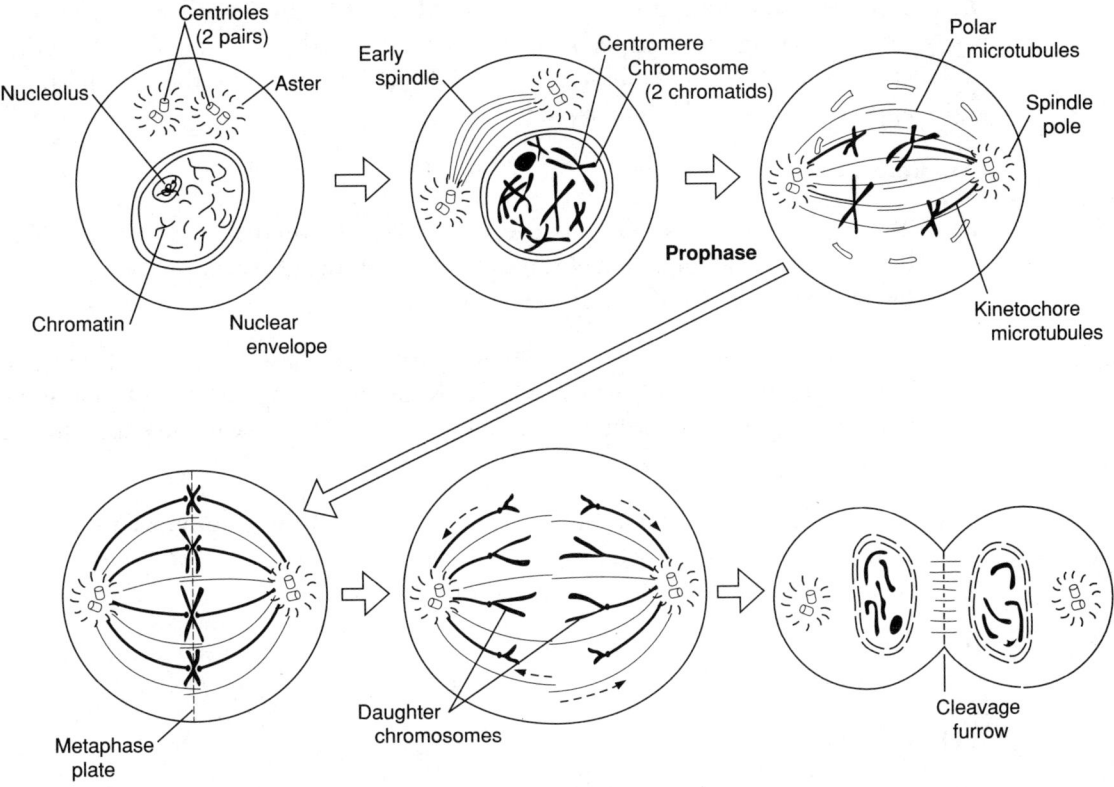

Figure 6-10. The process of mitosis. Note that in anaphase, microtubules shorten and slide apart, which separates daughter chromosomes.

microtubule-organizing center. The centrosome is initially composed of bundles of microtubules plus associated proteins assembled around the two pairs of **centrioles.** The centrosome then splits, forming two star-like microtubule bundles called **asters** surrounding each pair of centrioles.

2. Late prophase

   a. The **nuclear envelope breaks into membrane vesicles** that look like bits of ER.

   b. The **microtubules** in each aster **elongate,** moving the centrosomes toward opposite poles of the cell. The spindle is then bipolar, each consisting of two microtubule components. The **polar microtubules** extend from each pole toward the cell equator. The **kinetochore microtubules** extend from each pole and attach to the kinetochore, which is a special structure associated with each sister chromatid at the centromere. The kinetochores of sister chromatids attach to kinetochore microtubules extending from opposite poles.

3. Metaphase

   a. The **chromosomes align along the metaphase plate** (the equator), with balanced forces from opposite kinetochore microtubules. The sister chromatids of each chromosome are held together at the centromere.

    b. Each **chromosome is oriented perpendicular to the spindle,** with the kinetochores facing opposite poles. Each sister chromatid faces one pole or the other at random. When metaphase is seen from a polar viewpoint, the chromosomes appear to be spread out on a plate.

4. **Anaphase**

    a. The **centromere disconnects the two sister chromatids,** which separate to become identical daughter chromosomes. Each daughter chromosome moves the centromere first toward its respective pole.

    b. One force for chromosome movement during anaphase is the **shortening of kinetochore microtubules at the chromosome end.** As the chromosome maintains its attachment just ahead of the shortening microtubule end, it moves toward the pole.

    c. Another force for chromosome movement is the **sliding apart of polar microtubules extending from opposite poles.** This moves the poles further apart, thereby separating the chromosomes.

    d. At the end of anaphase, each pole has the same set of chromosomes (now single DNA molecules).

5. **Telophase**

    a. **Daughter nuclei are reformed** from the fragments of the nuclear envelope and other membrane vesicles of the parent cell.

    b. **The nucleoli reappear** and **the chromosomes condense** back into chromatin.

    c. Most cells undergo **cytokinesis** at the end of telophase. This process **divides the cytoplasm,** thereby partitioning nuclei and organelles into two separate cells. In animals, contraction of a ring of microfilaments forms the **cleavage furrow** and eventually pinches the cell in half. (In plants, convergence of vesicles forms a cell plate, which gives each daughter a plasma membrane. Substances are secreted between the membranes to form the new cell walls.)

    d. **If mitosis occurs without cytokinesis,** the result is a cell with **multiple nuclei** in the same cytoplasm. Such cells are characteristic of muscle tissue, slime molds, and filamentous fungi.

Specialized Eukaryotic Cells and Tissues

I. Overview

A. Specialization

During animal development, cells become specialized in structure and function (see Chapter 18). Individual cells are the basic unit for hierarchies of higher order structures.

B. Tissues

Tissues are groups of cells with a common structure and function. Each tissue also has its own distinctive characteristics (Table 7-1). This chapter describes the specialized cell types that make up various tissues in the animal body.

1. **Organs** are different tissues organized into specialized centers of function.

2. **Organ systems,** which are present in higher order animals, involve several organs that perform a major body function together. All the organ systems must be coordinated for the organism to survive.

II. Neural Cells and Tissues

A. Nervous System

Two internal communication systems coordinate the activities of the specialized parts of an animal—the nervous system and the endocrine system.

B. Cells of the Nervous System

The cells of the nervous system are specialized for quick reactions and direct control of target cells. These cells lay the foundation for building the highly ordered and complex structures of the nervous system. There are **two main classes of cells** in the nervous system—**neurons** and **supporting cells** (Figure 7-1).

1. **Neurons** transmit signals along the pathways of the nervous system. Neurons are the fundamental units of the nervous system, and all neurons have certain structures in common.

    a. The large **cell body** of the neuron contains the nucleus and most of the cytoplasm and organelles.

    b. **Processes** are long, fiber-like extensions from the cell body that enable the neuron to conduct signals over long distances. There are two types of neuron processes—**dendrites** and **axons.**

    (1) **Dendrites** are short, branched processes that transmit signals toward the cell body. Most neurons have numerous dendrites.

TABLE 7-1. Summary of Specialized Tissues

Tissue	Structure	General Functions	Examples
Nerve	Neurons with cell body, axons, dendrites	Sense stimuli, conduct impulses	Brain, spinal cord, peripheral nerves
Muscle	Long cells with actin and myosin microfilaments	Contract, move	Skeletal, visceral, cardiac tissue
Epithelial	Packed cells with basement membrane; cuboidal, columnar, squamous; simple or stratified	Protect, absorb, secrete, line body surfaces	Mucous membranes, glands
Connective	Few cells, secrete extra-cellular matrix with fibers	Connect and support other tissues	Loose, adipose, fibrous tissue; cartilage, bone, blood

    (2) **Axons** are long processes that transmit signals away from the cell body. Most neurons have only a single axon, extending from a part of the cell body called the **axon hillock.**

        (a) **Telodendria** and the **synaptic knob.** An axon may be branched, with each branch ending in numerous branchlets called telodendria. A synaptic knob is a bulbous structure at the tip of each telodendria.

        (b) **Neurotransmitters.** Nervous signals are transmitted by chemicals released from the synaptic knobs, known as neurotransmitters.

        (c) **Synapse.** The gap between the synaptic knob and the dendrite or cell body that receives the signal is the synapse. Neurotransmitters diffuse across the synapse. Communication between two neurons in a neuronal pathway occurs at the synapse between the cells.

2. **Supporting cells** provide structural support, protection, and assistance to neurons. There are many more supporting cells than there are neurons.

    a. **Schwann cells** are supporting cells arranged in chains along the axons of many neurons. Together, Schwann cells coat the axon with an insulating layer called the **myelin sheath.**

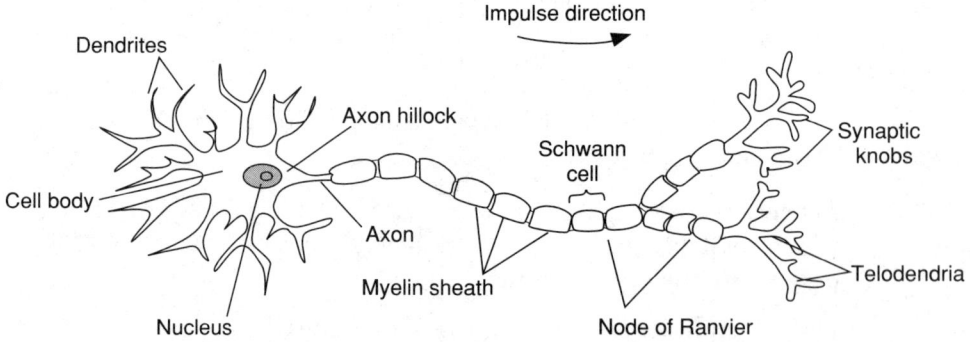

Figure 7-1. The structure of a myelinated neuron. Note the direction of impulse transmission.

b. The **nodes of Ranvier** are small gaps in the myelin sheath that occur between the serially arranged Schwann cells along the axon.

C. **Transmission Along the Neuron**

1. The **impulse transmitted along a neuron** depends on ionic changes (currents) across the plasma membrane of the cell. The cytoplasm of most cells is negatively charged relative to the extracellular fluid (see Chapter 6). Because of this ionic difference (voltage), the cell membrane has an electrical potential or **membrane potential**.

2. **Resting potential** is the membrane potential of a neuron that is not conducting an impulse.

   a. Most neurons have a resting potential of approximately **−70 mV.** The negative value indicates that the inside of the cell is negative relative to the outside.

   b. The resting potential results from the combined effects of **ion pumping** by the sodium–potassium pumps in the membrane and by **diffusion of ions** down their electrochemical gradients.

      (1) Ion pumping generates and maintains steep gradients of ion concentration, with sodium ions more concentrated on the outside and potassium ions more concentrated on the inside of the cell.

      (2) These ions can diffuse down their gradients through ion channels in the membrane: sodium ions tend to reenter the cell, and potassium ions tend to exit the cell.

      (3) The membrane is more permeable to potassium ions than to sodium ions, but the force on sodium ions is greater because they have an electrostatic attraction to the negative charge inside the cell.

3. **Graded potentials** are local changes in voltage induced by a stimulus. Such a stimulus might be physical pressure on the cell, an abrupt chemical change in the extracellular fluid, an electric shock, or a change in temperature. When a stimulus causes a local change in ionic differences across the membrane, it alters the membrane potential of the cell and affects the permeability of the plasma membrane. **Depolarization** and **hyperpolarization** are referred to as graded potentials. The strength of a graded potential decreases with distance from the point where the neuron was stimulated.

   a. **Depolarization** occurs when the resting potential becomes less negative, for example, from −70 to −50 mV, and the inside of the cell becomes more positive. Depolarization increases the chances that a nerve impulse will be triggered.

   b. **Hyperpolarization** occurs when the resting potential becomes more negative. In other words, the inside of the cell becomes more negative in relation to the outside of the cell. Hyperpolarization decreases the chances that a nerve impulse will be triggered.

4. **Action potential.** Neurons transmit ("fire") electrical impulses when graded potentials are converted to a larger depolarization called an **action potential** (Figure 7-2).

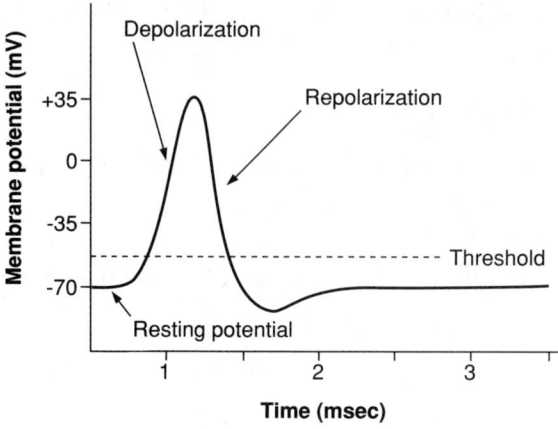

Figure 7-2. An action potential. Note the upsloping depolarization and the downsloping repolarization.

An action potential occurs near the point of stimulation and makes the inside of the cell positive compared with the outside (e.g., a change from −70 to +35 mV).

- a. **Voltage-sensitive gates** are membrane channels through which ions flow to generate the action potential. These channels open and close in response to changes in the membrane potential.

    (1) **Stimulus causes depolarization; amplification is possible.** A stimulus causes a local increase in the permeability of the membrane to sodium ions, which flow into the cell and depolarize the membrane. If this depolarization is large enough, the voltage-sensitive sodium ion gates open, increasing the permeability further. This amplification of a small depolarization generates the larger change in voltage of the action potential.

    (2) **Gates close; refractory period follows.** An action potential is a local and transient event because the sodium ion gates close quickly. After an action potential is triggered, there is a period of insensitivity to depolarization, during which time the sodium ion gates are inactivated. This refractory period lasts until the membrane is returned to its resting potential by repolarization of the membrane.

- b. **The membrane is repolarized** when the potassium ion gates open as the sodium ion gates close. Flow of potassium ions out of the cell makes the membrane potential more negative, even to the point of hyperpolarization, before the resting potential is restored.

- c. An action potential is triggered only when a stimulus is intense enough to cause a certain minimum depolarization (the **threshold potential**). A greater frequency of action potentials (i.e., repeated "firing") results from strong stimuli rather than from weak stimuli, but the magnitude of depolarization is the same.

5. **Propagation of the action potential.** The **axon hillock** is the zone where action potentials are usually generated. Thus, the chance of triggering an action potential depends on the distance from the point of stimulation (usually the dendrites or cell body) to the axon hillock. An action potential is transmitted, or propagated, from the axon hillock along the length of the axon (Figure 7-3).

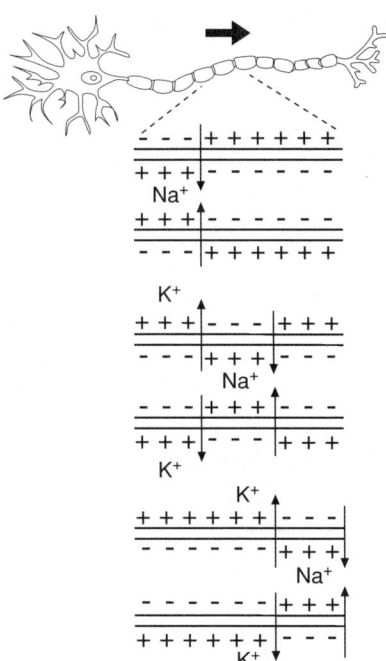

Figure 7-3. The sodium and potassium ion flows associated with a propagating action potential. Note the influx of sodium into the cell and the outflow of potassium from the cell.

    a. **Nerve impulse.** When an initial action potential occurs, sodium ions flowing into the cell also diffuse laterally, triggering an action potential at a site a little farther along the axon. Thus, the action potential is transmitted along the axon. The nerve impulse is the self-propagating wave of depolarization.

    b. **Serial transmission.** Nerve impulses are serially transmitted in one direction along the axon. Just behind a local depolarization, the refractory period from the previous action potential prevents a new one from being triggered.

    c. **Saltatory conduction.** Instead of being propagated continuously over the length of the axon, an action potential **"jumps" from one node of Ranvier to the next,** skipping the insulated regions in between. This saltatory conduction results in faster transmission of the nerve impulse. Saltatory conduction occurs because the nodes of Ranvier are where the voltage-sensitive ion channels are concentrated and where ions in the extracellular fluid are in direct contact with the neuron membrane.

D. Transmission Across the Synapse

    1. **Presynaptic and postsynaptic cells.** Signals are transmitted between two neurons in a pathway at the synapse between the two cells. The cell **transmitting** the signal is called the presynaptic cell and the cell **receiving** the signal is called the postsynaptic cell.

    2. **Types of synapses.** There are two types of synapses, electrical and chemical. Chemical synapses are much more common.

        a. **Electrical synapses** allow action potentials to spread directly from the presynaptic cell to the postsynaptic cell. The cells are connected by gap junctions, which are intercellular channels that allow ions to flow between the cells. Because of this electrical coupling, an impulse can be transmitted quickly from neuron to neuron.

  b. **Chemical synapses** (Figure 7-4) are narrow gaps (i.e., **synaptic clefts**) between the presynaptic cell and postsynaptic cell. The presynaptic cell converts the electrical signal of the action potential into a chemical signal that travels across the synapse to the postsynaptic cell. There, the signal is converted back into an electrical signal. Nerve impulses are transmitted in only one direction at a chemical synapse.

  (1) **Synaptic vesicles** exist within the synaptic knob of the presynaptic cell. The synaptic vesicles contain thousands of molecules of neurotransmitters, the chemical intercellular messenger in the synaptic cleft. One of the most common neurotransmitters is **acetylcholine (ACh),** which stimulates contractions at synapses between motor neurons and skeletal muscle cells.

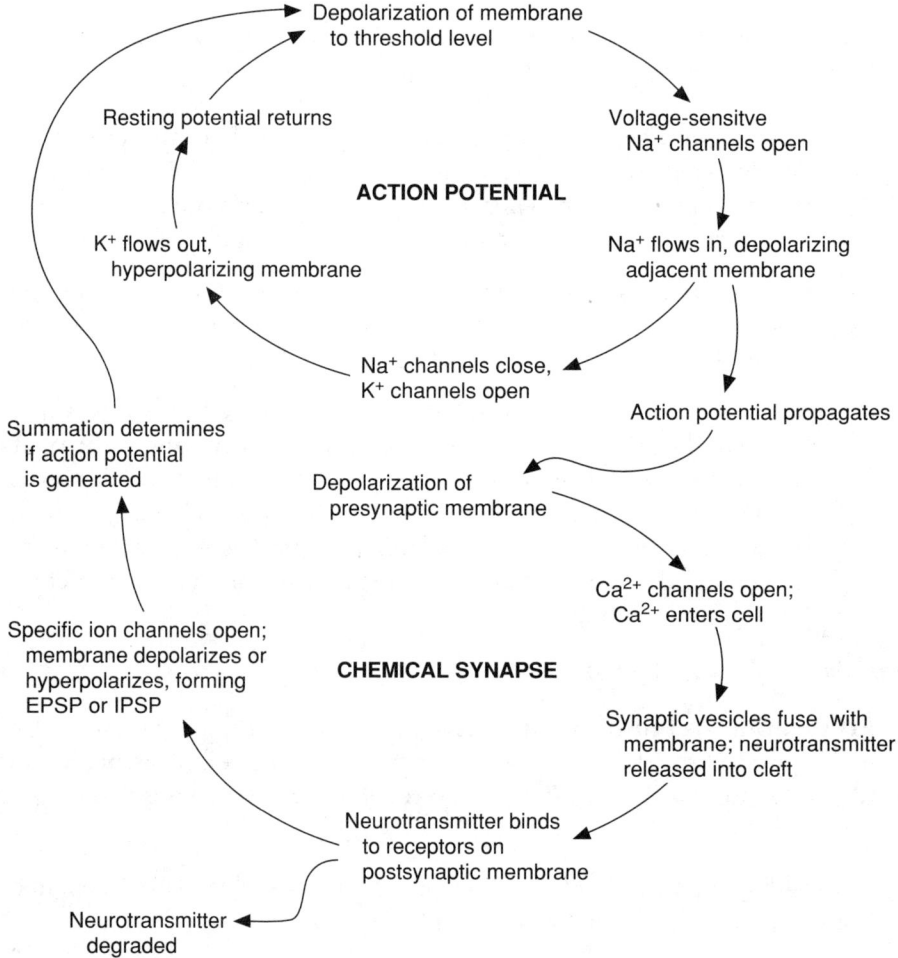

Figure 7-4. Summary diagram showing the flow of an action potential. *EPSP* = excitatory postsynaptic potential; IPSP = inhibitory postsynaptic potential.

(2) **The presynaptic membrane is depolarized** when an action potential being propagated along an axon reaches the synaptic knob. Calcium ions rush into the cell through voltage-sensitive calcium ion channels, stimulating the synaptic vesicles to fuse with the presynaptic membrane. The **neurotransmitter is released by exocytosis** into the synaptic cleft, where it diffuses to the postsynaptic membrane.

(3) The postsynaptic membrane has **neurotransmitter receptors** for specific neurotransmitter molecules. Binding of a neurotransmitter to its receptor opens the ion channels associated with the receptor. The resulting flow of a particular ion (e.g., Na^+, K^+, or Ca^{2+}) alters the membrane potential.

3. **Types of receptors.** Whether the membrane is depolarized (excited) or hyperpolarized (inhibited) by neurotransmitter binding depends on the type of receptors and the ion gates they control. **Excitatory** and **inhibitory signals** from many neurons often converge on a single neuron. The synapse is where all this information is integrated.

   a. At an **excitatory synapse,** the open gates allow flow of sodium ions into the cell, which depolarizes the membrane. This electrical change is called an excitatory postsynaptic potential (EPSP).

   b. At an **inhibitory synapse,** the open gates allow flow of potassium ion out of the cell (or chloride ion into the cell), which hyperpolarizes the membrane. This change is called an inhibitory postsynaptic potential (IPSP).

4. **Neurotransmitter degradation.** Degradation of the released neurotransmitter quickly silences each signal at a chemical synapse. The smaller chemical components are recycled to the presynaptic cell for synthesis of more neurotransmitters.

5. **Summation.** Postsynaptic potentials have an additive effect at the axon hillock called summation. The process makes the postsynaptic potentials strong enough to depolarize the region of the axon hillock to threshold potential. This allows an action potential to be triggered. There are two mechanisms of summation.

   a. In **temporal summation,** synaptic knobs release neurotransmitters in rapid-fire succession.

   b. In **spatial summation,** several different synaptic knobs, usually from different presynaptic neurons, stimulate a postsynaptic cell at the same time.

III. Contractile Cells and Tissues

A. Muscle Tissue

1. **Contraction.** Animal movement involves contraction of muscles. This contractile ability is characteristic of the long, excitable cells that make up muscle tissue, the most abundant tissue in the body. Contraction is the only action of a muscle (extension occurs passively).

   a. **Actin and myosin** are the longitudinally arranged microfilaments that comprise muscle cells.

b. **All muscle cells require electrical stimulation** (action potentials) to contract.

2. **Types.** There are three types of muscle tissue that differ in structure and function: **skeletal, cardiac, and visceral (smooth) muscle.** The cells that make up these muscles have different innervating neurons, membranes, and electrical properties.

   a. **Skeletal muscle** is generally responsible for the **voluntary movements** of the body. Skeletal muscle is attached to the bones by tendons.

      (1) **Myofibrils.** A skeletal muscle is a bundle of long fibers that extends the entire length of the muscle. Each muscle fiber is a single long cell with many nuclei, composed of bundles of smaller myofibrils.

      (2) **Myofilaments.** The longitudinally arranged myofibrils are composed of two types of myofilaments.

         (a) **Thin filaments** consist of coils of two strands of **actin** and one strand of a regulatory protein, **tropomyosin.**

         (b) **Thick filaments** consist of linear arrays of **myosin** molecules.

      (3) **Sarcomeres.** Skeletal muscle is also called **striated** muscle because of its striped appearance under the light microscope. This repeating pattern of light and dark bonds results from the regular arrangement of thin and thick filaments along the length of the fiber. Each repeating unit is called a sarcomere. The special arrangement of myofilaments in the sarcomere is the basis for the ability of the muscle to contract.

   b. **Cardiac muscle** forms the contractile wall of the heart. Cardiac muscle is also striated, but, unlike skeletal muscle, the cells are branched.

      (1) **Intercalated disks.** Special gap junctions that electrically couple (interconnect) all the muscle cells of the heart are called intercalated disks. Thus, action potentials generated in one cell can spread to all the cells, causing the entire heart to contract.

      (2) **Involuntary contractions.** In contrast to other muscles, cardiac muscle cells do not require stimulation by neurons, but can generate action potentials on their own. These action potentials last a long time because of a relatively high permeability of the membrane to sodium. Thus, contractions of the heart last a long time, with long periods of relaxation (refractory period) in between.

   c. **Visceral (smooth) muscle** is generally responsible for **involuntary movements.** Visceral muscle is found in the walls of the digestive tract, bladder, arteries, and other internal organs. The cells have a distinctive spindle shape.

      (1) **Nonstriated.** Visceral muscle is also called **smooth muscle** because of its lack of cross striations. Rather than being aligned along the length of the cell, the myofilaments are arranged spirally within smooth muscle fibers.

(2) **Weaker contractions.** The contractions of smooth muscle are weaker than those of striated muscle because the actin and myosin are organized differently. However, smooth muscle has a much greater range of lengths over which the cells can contract. Smooth muscles contract more slowly than striated muscle, but they can retain a contraction for a greater period of time.

B. The Sarcomere: Structure and Mechanism of Contraction (Figure 7-5)

1. **Structure**

    a. **Z lines.** The striations in skeletal muscle result from the lining up of Z lines, which are the edges of each sarcomere, of numerous myofibrils.

    b. **I band.** The thin filaments are attached to the Z lines and extend inward toward the center of the sarcomere. The thick filaments are centered within the sarcomere. In a resting muscle, there is an area near the edges of the sarcomere where the thin and thick filaments do not overlap. This region of only thin filaments is the I band.

    c. **A band.** The broad central region where the thin and thick filaments overlap is the A band.

    d. **H zone.** The narrow region in the center of the A band where there are only thick filaments is the H zone. (The thin filaments extend only partially across the sarcomere.)

2. **Sliding filament model.** In the sliding filament model, which is the mechanism of muscle contraction, the thin filaments slide across the thick filaments to pull the Z lines together and shorten the sarcomere. The I bands shorten and the H zone disappears, but the length of the A bands does not change. Shortening of all the sarcomeres in a myofibril allows the muscle to contract to approximately half its resting length.

    a. **Actin and myosin interactions** allow the filaments within the sarcomere to slide by each other (Figure 7-6). The myosin molecules of the thick filaments

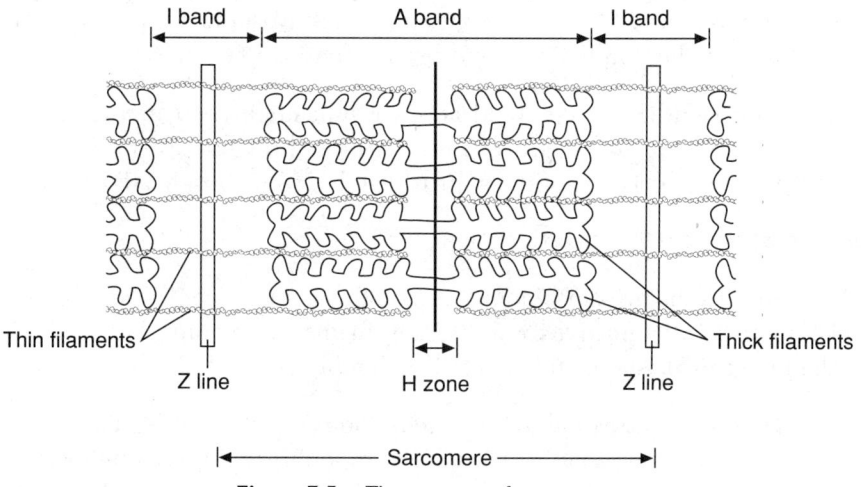

Figure 7-5. The structure of a sarcomere.

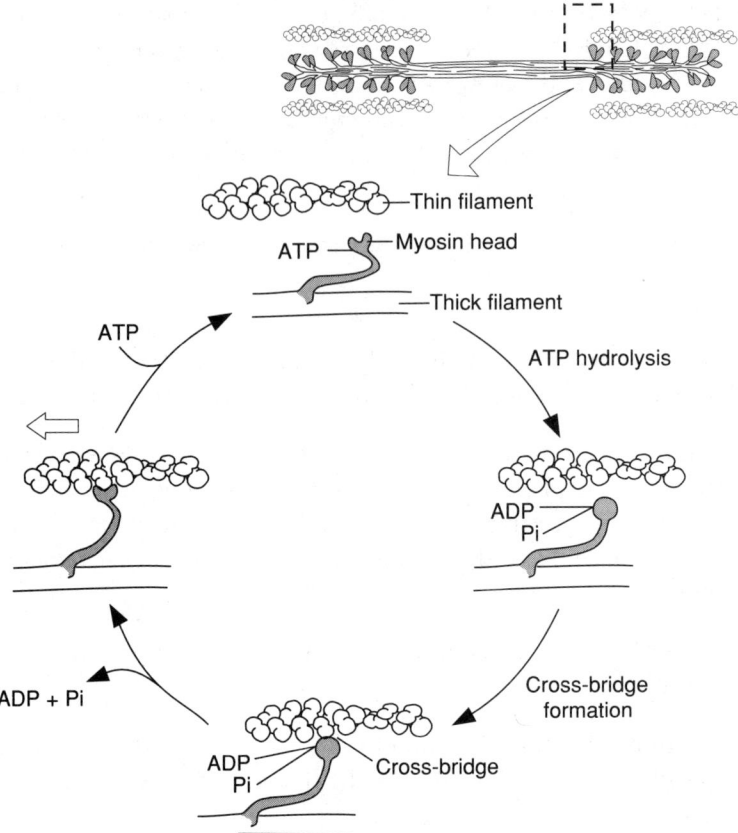

Figure 7-6. The mechanism of muscle contraction. Note the involvement of adenosine triphosphate (ATP) in this process. *ADP* = adenosine diphosphate.

        have "heads" that can bind to specific sites on the actin molecules of the thin filaments, resulting in **cross-bridges** between the thick and thin filaments. When cross-bridges form, the bent position of the myosin head pulls the thin filament toward the center of the sarcomere.

    b. **Cross-bridges can be broken only with an input of energy.** Adenosine triphosphate (ATP) is hydrolyzed by the myosin head, which breaks its attachment to the thin filament and returns to its original position.

    c. **To repeat the cycle,** the free myosin head binds to another site farther along the thin filament. Approximately five cross-bridges are formed and reformed every second by each of the approximately 350 myosin heads of a thick filament.

C. Regulation of Muscle Contraction

1. Rest. At rest, the myosin-binding sites on the actin molecules are blocked by the regulatory protein **tropomyosin** in the thin filament. A complex of regulatory proteins, called **troponin,** is also attached to the thin filament.

    a. **Troponin binds calcium.** Calcium ions are the key regulators of contraction. When troponin binds calcium, the troponin-tropomyosin interaction is altered, and the tropomyosin is displaced from the myosin-binding sites. Thus, muscle contraction occurs in the presence of calcium. When calcium levels drop, the myosin-binding sites are covered again, and contraction stops.

- **b.** The **sarcoplasmic reticulum** is a membrane system that accumulates and releases calcium ions.

2. **Motor neurons.** A muscle contracts only when stimulated by a motor neuron. The coupling of electrical excitation with muscle contraction occurs as follows:

    - **a.** An **action potential** occurs in the motor neuron, innervating a muscle cell, and the neuron releases ACh into the neuromuscular junction. This results in a graded depolarization of the plasma membrane of the muscle cell. If the depolarization is sufficiently large, it triggers an action potential that spreads across the muscle cell membrane.

    - **b.** **Transverse (T) tubules** are infoldings of the plasma membrane that carry the action potential into the muscle cell. Where the T tubules contact the sarcoplasmic reticulum, the action potential depolarizes the sarcoplasmic reticulum membrane and stimulates it to release calcium ions. Binding of calcium to troponin allows the muscle to contract.

    - **c.** **To safeguard against excessive contraction,** the sarcoplasmic reticulum pumps calcium back out of the cytoplasm as soon as the action potential passes, preventing further contraction.

    - **d.** The **energy** for this process comes from **ATP,** which is produced by the transfer of phosphate from a molecule called creatine phosphate to adenosine diphosphate.

3. **Muscle fiber innervation**

    - **a.** Some **small muscles,** such as those controlling eye movements, require a fine degree of control. In such a case, each motor neuron innervates only one fiber.

    - **b.** In **larger muscles,** each motor neuron may innervate hundreds of fibers, which are scattered throughout the muscle.

    - **c.** A **motor unit** is a single motor neuron plus all the muscle fibers it controls.

4. **Graded response.** Muscle contraction is a graded response, meaning that the **extent and strength of contraction can be varied.** There are two mechanisms responsible for the graded response.

    - **a.** The **number of motor units varies.** The strength of a contraction depends on the number of motor units that are stimulated. **Recruitment** of additional motor units results in gradually stronger contraction.

    - **b.** The **rate of successive contractions varies.** Whereas a single stimulus results in a muscle twitch, two stimuli in succession result in two twitches, the second of which is stronger than the first. This **wave summation** occurs because the second contraction starts before the muscle has completely relaxed from the first contraction. Although the muscle is still shortened, it contracts even further. Because motor neurons usually transmit stimuli in rapid succession, wave succession results in smooth, continual muscle contraction, called **tetanus**.

IV. Epithelial Cells and Tissues

A. **Structure and Function**

1. **Function.** Epithelial tissue **covers the outside of the body** and **lines organs and body cavities** inside the body. Epithelium provides an effective **barrier** and **protects a surface** against invading microorganisms, water loss, and physical injury.

2. **Structure.** Epithelial tissue consists of **sheets of tightly packed cells.** Epithelial cells are joined closely, in many cases, by tight junctions between the cells. Little if any material can pass between epithelial cells. An epithelium has **two surfaces.**

    a. The cells at the **free surface** are exposed to air or fluid.

    b. The cells at the **base** of the barrier are attached to a **basement membrane.**

3. **Cell shapes.** One criterion for grouping different types of epithelia is the shape of the cells on the free surface. The different cell shapes are **cuboidal** (like dice), **columnar** (like bricks on their ends), and **squamous** (like flat tiles).

4. **Specialized epithelia.** In addition to protecting the organs that they line, some epithelia are specialized for absorbing or secreting fluids. These epithelia are usually made up of cells with large cytoplasmic volumes (i.e., columnar or cuboidal epithelia).

    a. **Absorbing.** Nutrients are absorbed by the columnar epithelial cells that line the stomach and small intestine.

    b. **Secreting.** Substances are secreted by the cuboidal epithelia of the thyroid gland and kidney tubules.

5. **Mucous membrane.** Another type of epithelium is the mucous membrane, which secretes a solution that keeps a surface moist and lubricated. The free surfaces of some mucous membranes have cilia that move the thin layer of mucus along the surface. In the respiratory tubes, for example, the ciliated epithelium sweeps dust and other particles back up the windpipe to help keep the lungs clean.

B. **Types** (Figure 7-7)

In addition to shape, epithelia are classified by **number of cell layers.** The cells at the free surface may be cuboidal, columnar, or squamous, regardless of whether the epithelia has one or more layers.

1. A **simple epithelium** has a **single layer** of cells.

    a. **Simple squamous epithelia** are relatively leaky, allowing materials to move across the cell by diffusion. Epithelia specialized for this function are found lining the blood vessels and air sacs of the lungs.

    b. **Simple columnar epithelia** are specialized for absorption or secretion.

2. A **stratified epithelium** has **multiple layers** of cells. This type of epithelium can regenerate relatively easily. As old cells are sloughed off the free surface, cells near the basement membrane divide and push new cells to the free surface. Such epithelia are found on surfaces that are continually exposed to abrasive conditions, such as the outer skin and the linings of certain body passages and cavities.

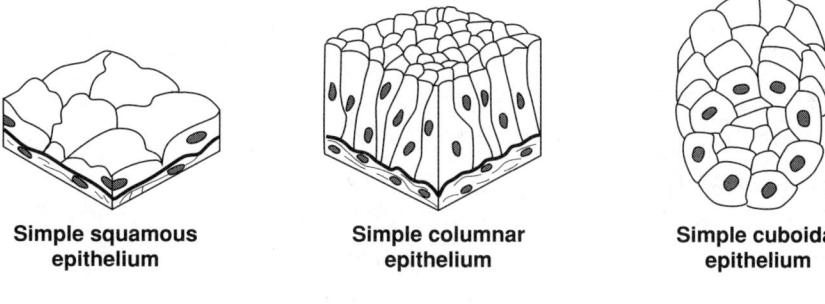
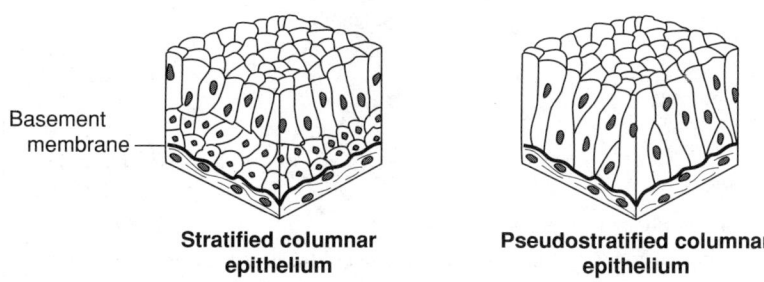

Figure 7-7. Various types of important epithelia, which are named based on cellular shape. Squamous means flat; columnar means column-like; cuboidal means cube-like ; and stratified means layered. Pseudostratified cells are in a unique layered structure.

3. A **pseudostratified epithelium** appears stratified because the cells vary in shape. It is actually a **single layer** of cells.

V. Connective Cells and Tissues

A. **Structure and Function** (Figure 7-8)

1. **Function.** Most connective tissues **bind and support other tissues.**

2. **Structure.** Unlike epithelia, connective tissues are composed of small numbers of cells dispersed in a matrix of extracellular material. This **extracellular matrix** is composed of a network of fibers in a homogeneous ground substance. The ground substance may be liquid, gel, or solid, depending on the type of connective tissue.

3. **Types**

   a. **Loose connective tissue** is the most abundant connective tissue, named for its loose web of fibers. Loose connective tissue **binds epithelia to underlying tissues and holds organs in place.** There are several kinds of loose connective tissue. Three types of loose connective tissue are distinguished by the proteins that make up the extracellular fibers. These fibers are secreted by cells called **fibroblasts.**

      (1) **Collagenous fibers** are bundles of fibrils, each of which is made of a coiled arrangement of the protein collagen. A distinctive property of collagenous fibers is their great strength along the length of the fiber (**ten-**

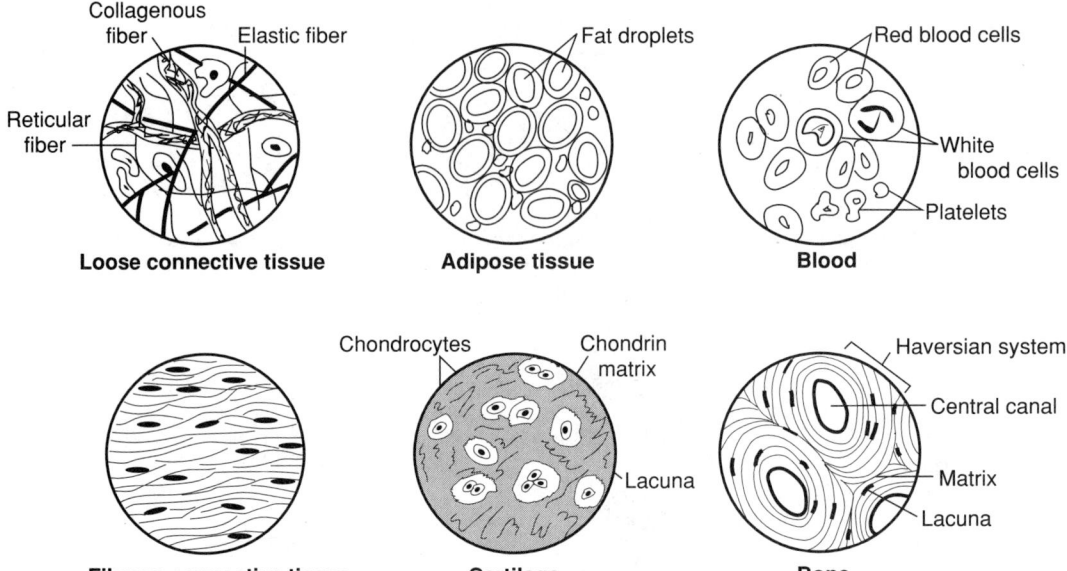

Figure 7-8. The microscopic appearance of various tissue types.

sile strength). These fibers are resistant to stretching, which is important for holding body parts together (e.g., keeping skin attached to bone).

- (2) **Elastic fibers** are long thread-like fibers made of the protein **elastin.** These stretchy fibers give the connective tissue **great resilience** to insult (e.g., the elasticity of skin).

- (3) **Reticular fibers** are branched and form a tight network of fibers. Reticular fibers **join connective tissue to adjacent tissues.**

b. **Adipose cells** are specialized for fat storage. They are scattered throughout **adipose tissue,** which is a type of loose connective tissue. Adipose tissue provides padding and insulation for the body, as well as storage for fuel molecules. Each adipose cell swells when fat is stored and shrinks when the fat is used as fuel.

4. **Blood** is composed of several types of cells suspended in a liquid extracellular matrix called plasma. It differs from other connective tissues by its function in circulation, rather than in binding and support.

   a. **Plasma** consists of water, salts, and various dissolved proteins.

   b. **Three types of cells** are suspended in plasma.

   - (1) **Erythrocytes** (red blood cells) carry oxygen throughout the body.
   - (2) **Leukocytes** (white blood cells) function in the immune system.
   - (3) **Platelets** are involved in blood clotting.

5. **Fibrous connective tissue.** Dense or fibrous connective tissue has a greater proportion and degree of organization of collagenous fibers than loose connective tissue. The dense fibers are arranged in parallel bundles, which give them greater tensile

strength. The **tendons** that attach muscles to bones and the **ligaments** that join bones together at joints are both composed of dense connective tissue.

B. Cartilage and Bone

1. **Cartilage** is a dense mesh of collagenous fibers in a ground substance called **chondrin,** which is a protein-carbohydrate complex with a rubbery consistency. Chondrin is secreted by cells called **chondrocytes,** which are localized in spaces called lacunae that are scattered through the ground substance. The combination of collagenous fibers and chondrin makes cartilage a strong yet flexible material. The main function of cartilage is **support and reinforcement** (e.g., for the nose, ears, vertebral disks).

2. **Bone,** which is a mineralized connective tissue, forms the skeleton that supports the body of most vertebrates.

    a. **Osteocytes** are cells that secrete a collagenous matrix and release calcium phosphate, which forms deposits of the mineral **hydroxyapatite** within the matrix. Bone is hardened without becoming brittle by this combination of hard mineral and flexible collagen.

    b. **Haversian systems** are the repeating units that make up hard bone. Each system is composed of **lamellae,** which are concentric layers of the mineralized matrix. The lamellae are deposited around a central area that contains blood vessels and nerves that provide materials to the bone.

    c. **Lacunae** are the small spaces that contain osteocytes. The lacunae are surrounded by the hard matrix and are interconnected by cellular extensions called **canaliculi.**

    d. **Marrow** is spongy bone tissue that forms the interior of long bones. The long bones have a hard outer region made of Haversian systems. The **red marrow** near the ends of long bones is where blood cells are produced.

SECTION II

Physiology

The Nervous System

I. Introduction

A. Function

The human nervous system allows the body to respond to its environment by monitoring sensory stimuli, interpreting the information provided by the senses, and suggesting an appropriate response. The mechanism of the nervous system is complex, with multiple levels of regulation and millions of interconnecting networks. Therefore, when studying this complex system, it is important to understand the anatomic organization of the nervous system with all of its multiple subdivisions (Figure 8-1).

B. Structure

The human nervous system has two anatomic subdivisions. The **peripheral nervous system (PNS)** includes both the cranial nerves and the spinal nerves, as well as their associated ganglia. The **central nervous system (CNS)** is composed of the brain and the spinal cord.

II. Peripheral Nervous System

A. Structure

1. There are **12 pairs of cranial nerves** (Table 8-1) that exit the brain directly; there are **31 pairs of spinal nerves** that exit from the spinal cord.

2. The **spinal, or peripheral, nerves** exit through the vertebral column through the intervertebral foramina and are named according to the vertebral level from which they exit.

3. The cell bodies of spinal nerve sensory fibers are located in the **dorsal root ganglia** (Figure 8-2), and those of motor fibers are present within the **ventral horns.**

4. Normally, **sensory impulses travel to the brain,** where the sensory impulses are processed and motor impulses send information to the periphery to dictate the desired response.

5. In the case of a **reflex arc,** the CNS is bypassed. In the simplest scenario, a sensory receptor stimulates a sensory nerve, which then synapses either directly or indirectly (via interneurons) to a motor nerve. The motor nerve then stimulates the effector tissue, completing the circuit.

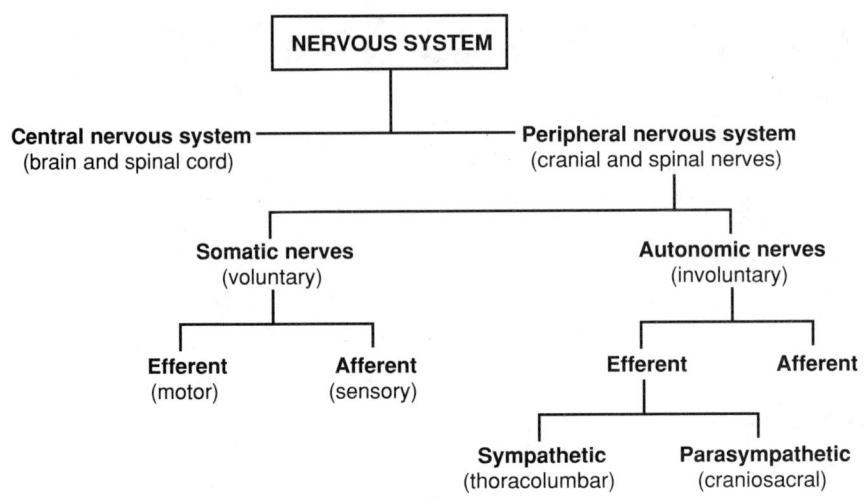

Figure 8-1. A flow diagram of the nervous system.

TABLE 8-1. The Cranial Nerves

Number	Cranial Nerve	Components	Innervation and Function
I	Olfactory	Sensory	Nasal epithelium (olfaction)
II	Optic	Sensory	Rods and cones of retina (vision)
III	Oculomotor	Motor	Eye muscles (movement of eye); pupil constriction; ciliary muscle (accommodation)
IV	Trochlear	Motor	Eye muscle (movement of eye)
V	Trigeminal	Sensory	Skin of face and mucosal surfaces of mouth and nose (sensation)
		Motor	Muscles of mastication (chewing)
VI	Abducens	Motor	Eye muscle (movement of eye)
VII	Facial	Sensory	Taste for the anterior two thirds of the tongue
		Motor	Muscles of facial expression
VIII	Vestibulocochlear	Sensory	Cristae of semicircular canals (balance and equilibrium); hair cells of organ of Corti (hearing)
IX	Glossopharyngeal	Sensory	Taste for posterior one third of the tongue; sensation for pharynx
		Motor	Parotid gland (secretion)
X	Vagus	Motor	Constrictors of pharynx (swallowing); muscles of larynx (phonation); heart smooth muscle, and glands of gastrointestinal tract and pulmonary system (autonomic control)
XI	Spinal Accessory	Motor	Trapezius and sternomastoid muscles (elevation of shoulders and movement of head)
XII	Hypoglossal	Motor	Muscles of tongue (movement of tongue)

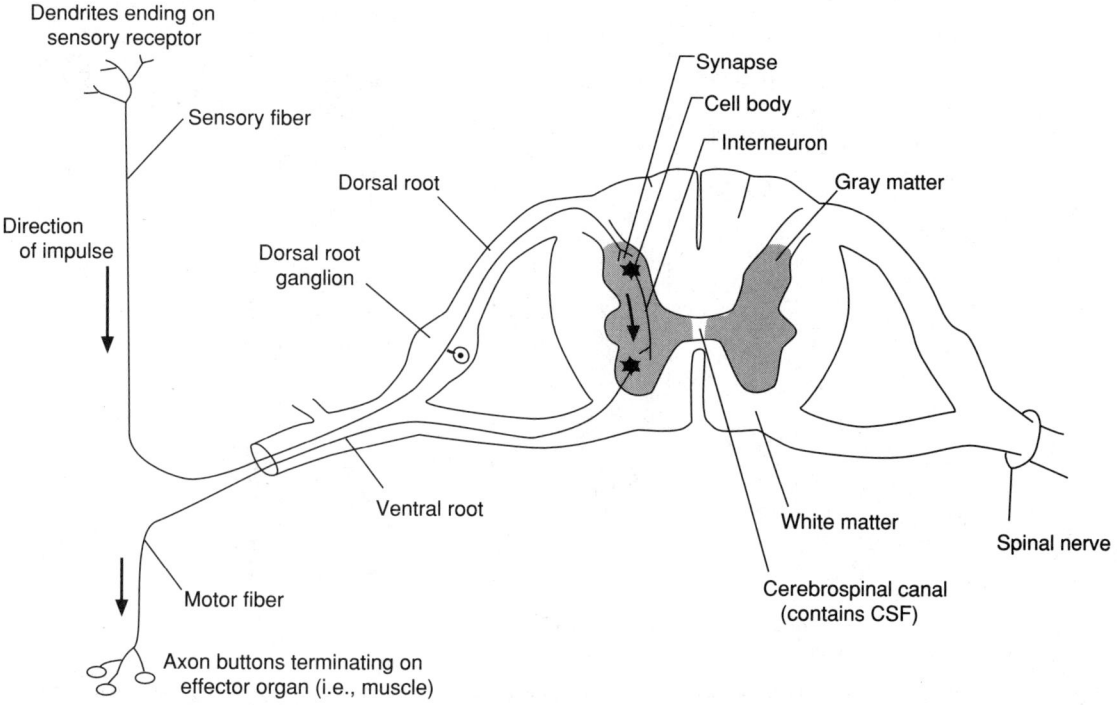

Figure 8-2. Cross section of a spinal cord with a reflex arc. *CSF* = cerebrospinal fluid.

B. Function

1. **Voluntary and involuntary functions**

    a. The **voluntary** components of the PNS are the **somatic nerves,** which innervate skeletal muscle.

    b. The **involuntary** components are the **autonomic nerves,** which are grouped together under a functional unit also known as the **autonomic nervous system (ANS).**

2. **Efferent and afferent nerve networks** are contained in both the somatic and autonomic branches of the PNS.

    a. The **efferent,** or **motor,** nerves carry nerve impulses from the CNS into the body.

    b. In contrast, the **afferent,** or **sensory,** nerves monitor sensory inputs and carry this information from the periphery to the CNS.

3. The **ANS** is involved in controlling the maintenance of **internal homeostasis.** The ANS **innervates all smooth muscle, cardiac muscle, and glandular tissue.**

4. The **motor division of the ANS** is divided once again into sympathetic (thoracolumbar) and **parasympathetic** (craniosacral) components (Figure 8-3). The **nerves** that comprise these two components have important anatomic distinctions.

    a. **Sympathetic nerves** arise from the **thoracic and lumbar regions** of the spinal cord.

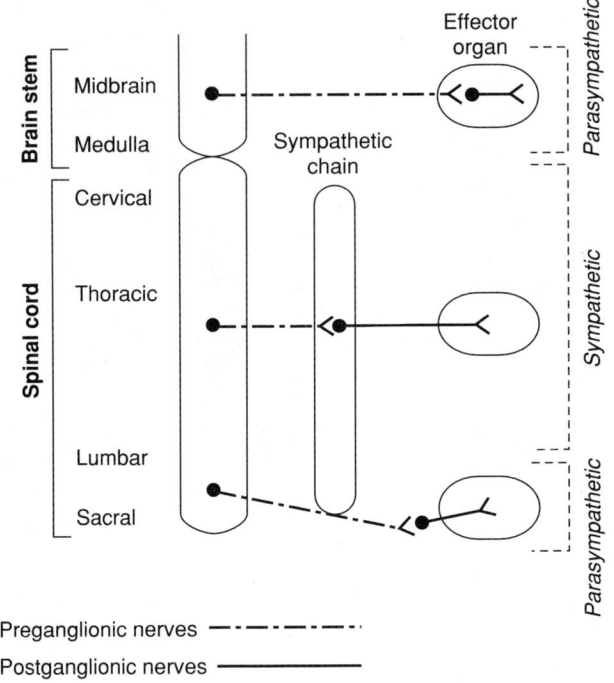

Figure 8-3. The sympathetic and parasympathetic nervous systems. Parasympathetic fibers arise in the brain stem or the sacral spinal cord, have a synapse in a ganglion close to the target, and finally synapse in the target organ or tissue. Sympathetic fibers originate in the thoracic or lumbar spinal cord, synapse nearby in the sympathetic chain, and finally synapse a distance away in the target organ or tissue. *CSF* = cerebrospinal fluid.

      b. **Parasympathetic nerve fibers** originate in the **cranial and sacral regions.**

      c. There are differences in the location of the ganglia where the sympathetic and parasympathetic nerves synapse.

5. A **ganglion** is a collection of nerve cell bodies that usually contains multiple synapses.

      a. In the **sympathetic** nervous system, chains of ganglia are found on **either side of the spinal cord.**

      b. In the **parasympathetic** nervous system, the ganglia are located **in or near the innervated organs.**

      c. As a consequence of this differing ganglial placement, a **third distinction** is made between the two systems.

           (1) **Efferent autonomic circuits** are composed of two nerves. **Preganglionic nerves** arise from the CNS and synapse in the ganglion. **Postganglionic nerves** then complete the circuit, carrying the impulse from the ganglion to the target tissue.

           (2) **Sympathetic circuits** have short preganglionic and long postganglionic fibers. **Parasympathetic pathways are the opposite,** having long preganglionic and short postganglionic nerve fibers.

TABLE 8-2. Effects of Autonomic Innervation on Effector Organs

Effector Organ	Parasympathetic Action	Sympathetic Action
Heart	Decreases rate; no effect on strength of contraction	Increases the rate and the strength of contraction
Blood vessels	Dilates	Constricts
Bronchial tubes	Constricts	Dilates
Pupils	Constricts	Dilates
Bladder	Constricts	Relaxes
Kidney	Increases urine production	Decreases urine production
Gastrointestinal tract	Stimulates smooth muscle contraction and secretion of enzymes	Inhibits smooth muscle contraction and gastrointestinal motility
Sexual function	Dilates blood vessels; erection	Constricts blood vessels; ejaculation
Salivary glands	Stimulates thin, watery secretions	Stimulates thick, viscous secretions
Energy mobilization: liver and adipose tissue	No effect	Stimulates glycogenolysis and fatty acid release respectively
Neurotransmitter	Acetylcholine	Epinephrine and norepinephrine

    (a) **Sympathetic preganglionic nerves tend to synapse with many postganglionic nerves.** This arrangement is contrasted by few synaptic connections in the parasympathetic system.

    (b) Therefore, **the effects of sympathetic activation tend to be more widespread than those of parasympathetic stimulation,** in which the effects are more localized.

6. **The sympathetic and parasympathetic divisions of the ANS exert opposing effects on the body.** See Table 8-2 for a list of autonomic influences on the body.

   a. The **sympathetic nerves** prepare the body for **"fight or flight,"** raising blood pressure, increasing respiration, and reducing energy expelled on digestive activities.

   b. The **parasympathetic nerves** encourage homeostatic activities (**"rest and digest"**), such as decreasing blood pressure, reducing respiration, and stimulating the glandular secretions necessary for digestion.

III. The Central Nervous System

A. The Spinal Cord (see Figure 8-2)

1. In the **center of the spinal cord** is a butterfly-shaped mass composed of the cell bodies of the spinal cord neurons. This area is known collectively as the **gray matter** because the cell bodies stain gray whereas the surrounding axons remain white. (The white matter does not stain with the laboratory dye because of its high lipid content.)

2. The cell bodies that conduct **sensory functions** are located in the **dorsal horns.**

3. The cell bodies involved in **motor activities** are in the **ventral horns.**

4. Surrounding the gray matter is the **white matter,** which is composed primarily of myelinated nerve fibers. This arrangement of gray and white matter extends the length of the spinal cord, differing only with respect to the ratio of gray matter to white matter.

5. The fibers of the white matter travel through the spinal cord in **tracts, or columns.** Each tract contains groups of fibers that share a common function (e.g., pain sensation for the right index finger) in addition to "sharing" similar sites of origin and termination.

    a. These nerve fibers enter or exit the spinal cord via the **dorsal or ventral roots,** depending on whether they are sensory or motor nerves, respectively.

    b. The dorsal and ventral roots then join to form a **spinal nerve** that **carries both sensory and motor fibers.**

6. The **spinal cord is protected** by the surrounding **vertebral column** as well as by three connective tissue sheaths known as the **meninges** of the brain.

    a. From exterior to interior, the meninges are called the **dura mater,** the **arachnoid membrane,** and the **pia mater.**

    b. **Cerebrospinal fluid (CSF)** fills the subarachnoid space between the arachnoid membrane and pia mater, **cushioning the spinal cord** from shock.

    c. In some **diseases** (e.g., malaria), the meninges become inflamed (meningitis) and a sample of the CSF is needed to determine the appropriate therapy. To prevent injury to the spinal cord, lumbar punctures or spinal taps are performed below the second lumbar vertebra (L2). The spinal cord terminates below L2 into nerve roots termed the **cauda equina.**

B. **The Brain**

The human brain can be divided into distinct **anatomic regions** with different functional domains (Figure 8-4).

1. The **medulla oblongata,** which is continuous with the spinal cord, is at the **caudal end** of the brain stem. Major activities controlled in this region of the brain include regulation of heart rate, respiration, and blood pressure. The reflex centers that control blinking, coughing, sneezing, swallowing, and vomiting are also in the medulla oblongata.

2. The **pons** is a bulged area on the anterior surface of the brain stem, just superior to the medulla oblongata. The pons aids the medulla in the control of respiration. It relays information between the cerebral cortex and the cerebellum, and it contains the cell bodies that control facial expression, mastication, lacrimation, and salivation.

3. Traveling up the brain stem, the **midbrain** is the next area. The midbrain contains the **red nucleus** (integrates information concerning muscle tone and posture), the **superior colliculi** (mediates visual reflexes), and the **inferior colliculi** (mediates auditory reflexes). The midbrain also controls pupillary constriction and accommodation.

4. The **cerebellum** is a bihemispheric structure located posterior to the brain stem. It is

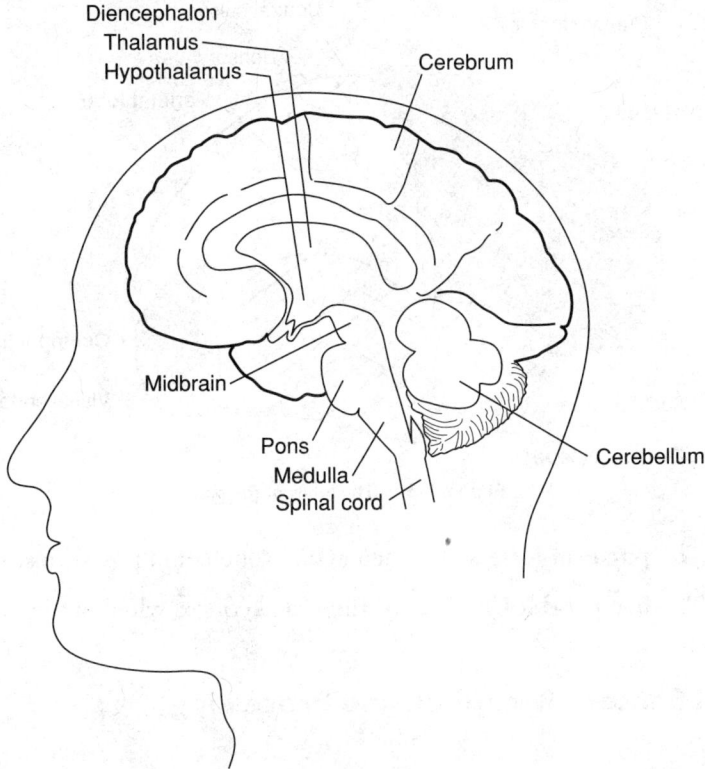

Figure 8-4. Structure of the human brain.

the second largest part of the brain and controls balance, maintenance of muscle tone and posture, and smooth, coordinated movement.

5. The **diencephalon** is the interior of the brain proper. There are three major components of this region.

    a. The **thalamus** acts as the main relay center, conducting information between the cerebral cortex and the rest of the nervous system.

    b. The **hypothalamus** controls body temperature, appetite, sleep, and blood pressure, it secretes releasing factors that regulate pituitary gland function, and it modulates some emotional and sexual responses (e.g., hostility, pain, pleasure). The hypothalamus is a member of the final component of the diencephalon known as the limbic system.

    c. The **limbic system** is concerned with emotion, including interpretation of emotional stimuli and behavioral responses to these stimuli.

6. The **cerebrum** is the largest and most prominent part of the brain. A central, longitudinal fissure divides the cerebrum into left and right hemispheres, which communicate with each other via the **corpus callosum.** Each hemisphere of the cerebrum is further subdivided into four main lobes (Figure 8-5).

    a. The **frontal lobe** controls voluntary movement **(motor cortex),** behavior, learning, thought, judgment, and personality.

    b. The **temporal lobe** interprets auditory stimuli and aids in spoken language.

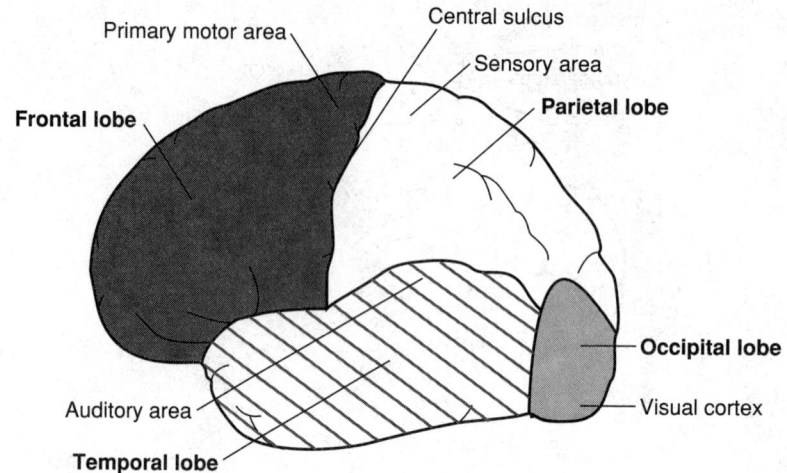

Figure 8-5. The lobes of the brain.

    c. The **parietal lobe** senses heat, cold, touch, and pressure stimuli.

    d. The **occipital lobe** contains the visual cortex, which interprets visual stimuli.

IV. Special Sensory Reception and Processing

A. Vision

1. **Structure** (Figure 8-6)

    a. The outermost layers of the eyeball, the **sclera** and the **cornea,** are composed of fibrous connective tissue.

    b. Just interior is a vascular, pigmented layer that makes up the **choroid,** the **ciliary body,** and the **iris.**

    c. The innermost layer—the **retina**—contains photoreceptor cells, namely the **rods** and **cones,** and other neural cells that aid in the transmission and integration of visual stimuli.

    d. The visual inputs are then carried via the **optic nerve** to the occipital lobe, where they are processed further.

2. **Mechanism of vision.** The mechanism of vision involves stimulation of the retina by light and the conversion of photon energy to electrical energy by photoreceptor cells.

    a. **Photons** of light travel through the cornea, the anterior chamber containing **aqueous humor,** the pupil, and ultimately the lens, where they become refracted, creating an inverse image.

    b. Once refracted, the photons pass through the **vitreous humor** until they reach the layers of the retina, causing stimulation of both rod and cone cells (photoreceptor cells).

    c. Most of the surface of the **retina** contains photoreceptors; however, there are **special regions.** The **optic disk,** the site where the optic nerve exits from the back of the eye, contains no photoreceptors, which creates an anatomic blind

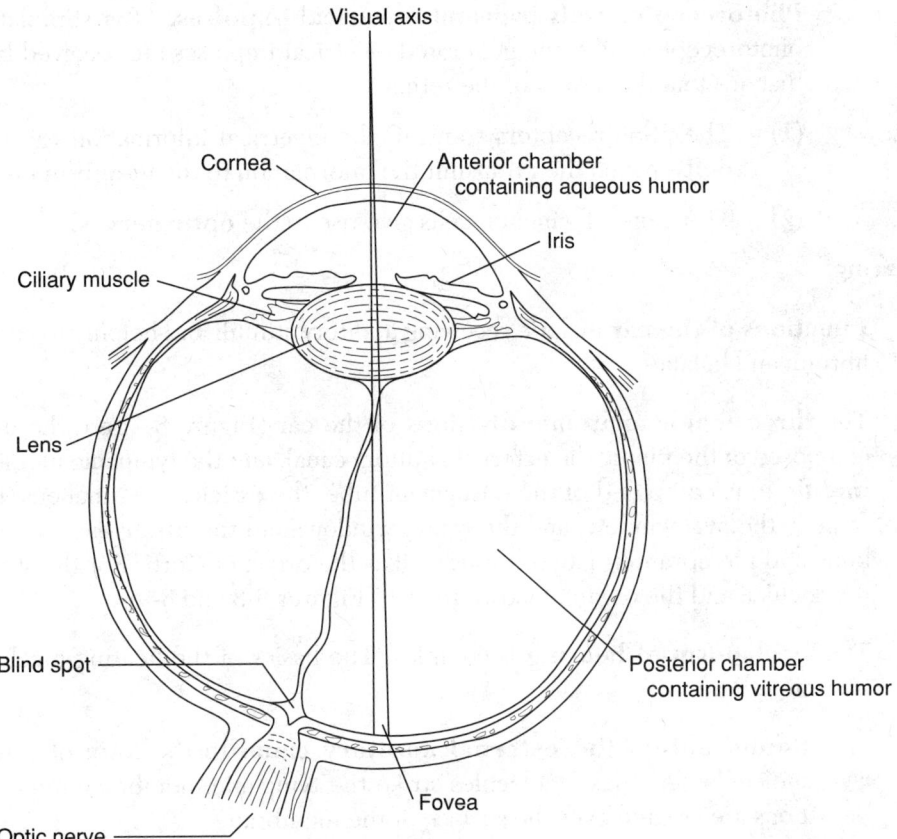

Figure 8-6. The structure of the eye.

spot. The area of highest visual acuity (i.e., it has the highest density of cone cells) is the **fovea.**

d. **Photopigment ultimately creates electrical energy.** When photons strike the photoreceptor cells, a photopigment absorbs the light and undergoes structural changes that cause the generation of ionic potentials. Therefore, photon energy is first converted into chemical energy and then into electrical energy.

   (1) Rod cells contain the photopigment **rhodopsin.** This photopigment contains a protein called opsin that is chemically joined to a vitamin A derivative called retinal. **Rods** differentiate light intensities (white/black) because they are sensitive and spread out over the surface of the retina. Thus, rods control vision in dim light (e.g., night vision).

   (2) **Cones** contain **iodopsin** as their photopigment.

   (a) There are **three types of cone cells,** each containing a different form of photopigment to discriminate red, green, and blue wave forms.

   (b) Because they are less sensitive than rod cells, cones enable **color perception and bright light vision.** Cones are best for good visual acuity function because they have great resolution ability.

   (c) Cones are **grouped together over the retinal surface,** which aids in their discrimination of fine detail.

e. **Photoreceptor cells transmit electrical impulses.** After stimulation of the photoreceptor cells, the generated electrical impulses are received by neurons that are found in layers of the retina.

   (1) The photoreceptors transmit the electrical information to the **bipolar cells,** which then transmit the information to the **ganglion cells.**

   (2) The axons of ganglion cells give rise to the **optic nerves.**

B. **Hearing**

   1. **Functions of the ear** include receiving auditory stimuli and helping to control equilibrium and balance.

   2. The **three major anatomic divisions** of the ear (Figure 8-7) are the **outer ear,** composed of the pinna, the external auditory canal, and the tympanic membrane; the **middle ear,** composed of the eustachian tube, the ossicles (i.e., malleus, incus, and stapes), the oval window, and the round window; and the **inner ear,** containing the bony and membranous labyrinths as well as the organ of Corti. For the structures of the cochlea and the organ of Corti, refer to Figures 8-8 and 8-9.

   3. The **mechanism of hearing** is complex. The basics of the hearing mechanism are presented here.

      a. **Sound enters the external auditory canal** in the form of waves of air molecules. As these molecules strike the tympanic membrane, unique oscillations are created over the surface of the membrane.

      b. **Oscillations of the tympanic membrane cause movement of the ossicles,** or middle ear bones, which amplify the signal and translate the displacement of the tympanic membrane into mechanical energy.

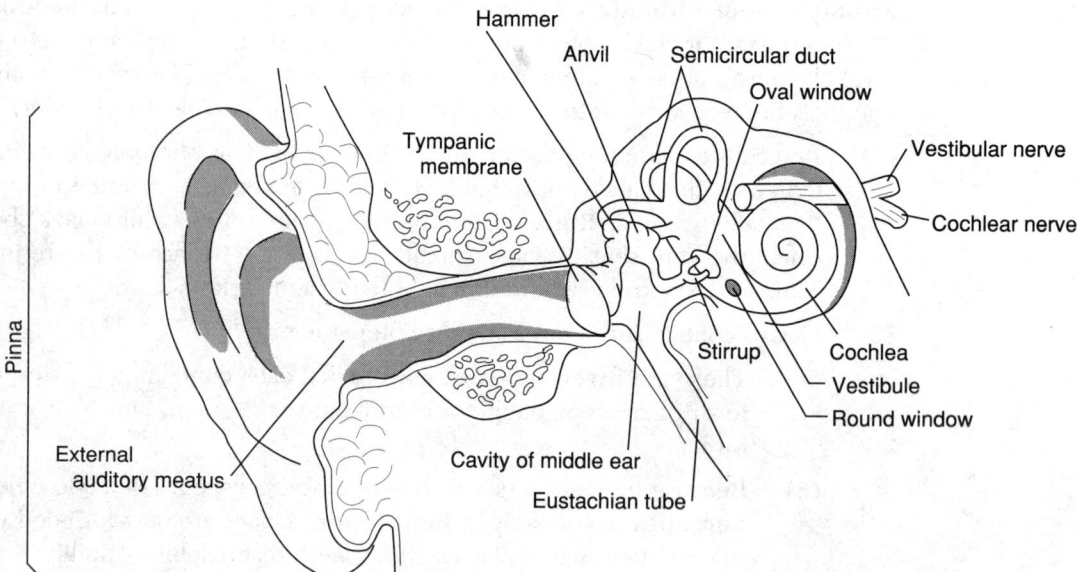

Figure 8-7. The structure of the human ear. (Reproduced with permission from Ville CA, Solomon EP, Davis PW: *Biology.* Philadelphia, Saunders College Publishing, 1985, p 887.)

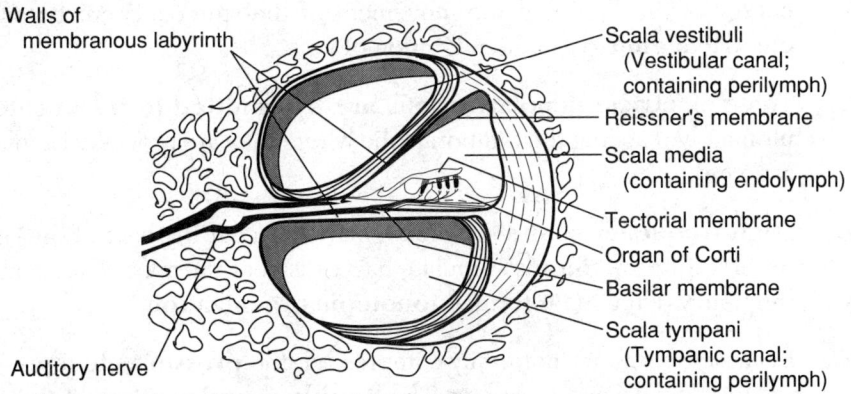

Figure 8-8. The cochlea. (Reproduced with permission from Ville CA, Solomon EP, Davis PW: *Biology*. Philadelphia, Saunders College Publishing, 1985, p 888.)

    c.   At the end of the three-bone chain (i.e., the ossicles), **the stapes covers the oval window of the cochlea,** which is a fluid-filled system of canals. When the stapes moves against the oval window, **perilymph** fluid within the **scala vestibuli (bony labyrinth)** of the cochlea creates wave forms of varying amplitudes. This generates pressure changes, which are translated to the **endolymph** of the **scala media (membranous labyrinth)** via distortions of the **Reissner membrane.**

    d.   Movement of the endolymph, varying with the pitch and volume of the eliciting sound, causes **movement of stereocilia** located on the hair cells of the organ of Corti.

    e.   Movement of the stereocilia is coupled with the **opening and closing of ion channels,** which, in turn, can lead to local depolarizations. The mechanical

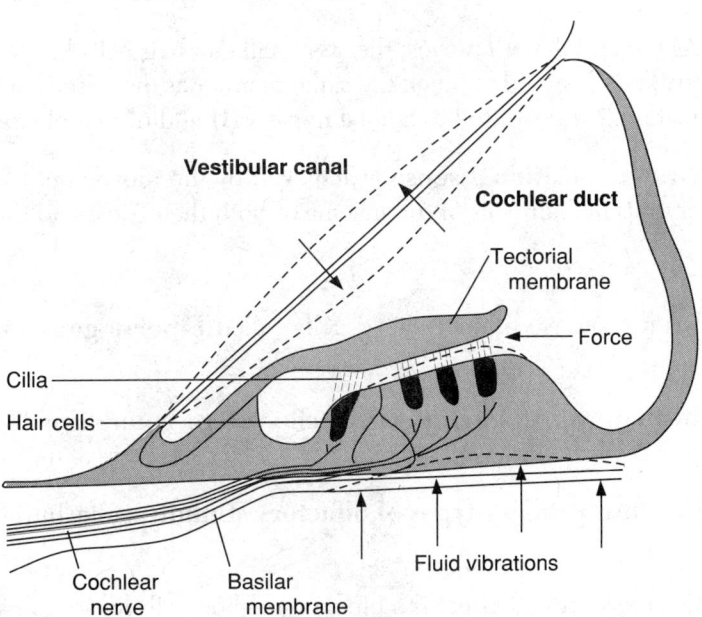

Figure 8-9. The organ of Corti. (Reproduced with permission from Ville CA, Solomon EP, Davis PW: *Biology*. Philadelphia, Saunders College Publishing, 1985, p 888.)

energy of the fluid and the movement of the stereocilia are **translated into electrical energy** via the migration of ions.

    f. These **electrical depolarizations are transmitted to the cochlear nerve,** ultimately reaching the temporal lobe where they are processed and interpreted as sound.

    g. **High-frequency sounds** stimulate hair cells near the base of the cochlea, and **low-frequency sounds** stimulate hair cells near the apex. This arrangement of frequency detection is termed **tonotopic organization.**

    h. Because liquids are not easily compressed, the **pressure changes** of the scala vestibuli need a way to escape. The **flexible round window at the end of the scala tympani** absorbs these pressure changes, acting as a pressure release valve. The pressure can either be **transferred directly,** because the scala vestibuli and scala tympani communicate at the apex at a point referred to as the **helicotrema,** or they can be **transmitted indirectly** from the endolymph via movement of the Reissner membrane and the basilar membrane.

4. In addition to its role in hearing, the ear also enables the body to control its position in space via a vestibular function controlling **balance.**

    a. The **vestibular organ** of the ear comprises the **three semicircular canals, the utricle, and the saccule.**

    b. Together, these canals monitor the effects of **gravity, body movement, and head position.**

    c. Displacement of the head causes **movement of endolymph within the semicircular canals.** Each canal responds to a different three-dimensional plane because the three canals are connected to the utricle at right angles to one another.

    d. As the endolymph moves, the stereocilia on hair cells located at the **crista ampullaris** respond in much the same manner as discussed for hearing. The information is transmitted to cranial nerve VIII and ultimately to the cerebellum.

    e. **Gravity positional sense** is achieved by the movement of calcium carbonate crystals on hair cells of the macula of both the utricle and the saccule.

C. Olfaction

1. **Smell** is a response of specialized cells called **bipolar ganglion cells** to gaseous stimuli.

2. The **mechanism of smell** is mediated via a receptor–ligand signal transduction mechanism.

3. There are **four primary types of olfactory stimuli:** acidic, burnt, fragrant, and rancid.

    a. Olfactory stimuli affect the bipolar ganglion cells located in a highly specialized mucous membrane in the **roof of each nasal cavity.**

b. The ganglion cells terminate in knobs that have several **olfactory hairs.** The hairs contain receptors that bind to the olfactory ligands. This creates **electrical impulses** inside the cells, which are then referred to cranial nerve I.

D. Taste

The mechanism of taste is analogous to that of smell.

1. **Neuroepithelial cells on taste buds** of the tongue act as receptors for substances in solution.

2. The **four types of taste receptors**—bitter, salt, sour, and sweet—are ligand-specific and are distributed over specific areas of the tongue.

   a. **Areas of the tongue.** Sour is perceived on the back part of the tongue. Salt and sour are perceived on the sides of the tongue, and sweet is perceived on the tip of the tongue.

   b. **Innervation and transmission.** Information regarding taste is relayed by cranial nerve VII for the anterior two thirds of the tongue, whereas the posterior one third is innervated by cranial nerve IX.

E. Touch

1. The **skin** functions in sensory reception; it contains nerve endings and specialized receptors that detect pain, touch, temperature, pressure, vibration, and proprioception (joint position sense in space).

2. **Free nerve endings** are stimulated directly by contact with an object on the surface of the body.

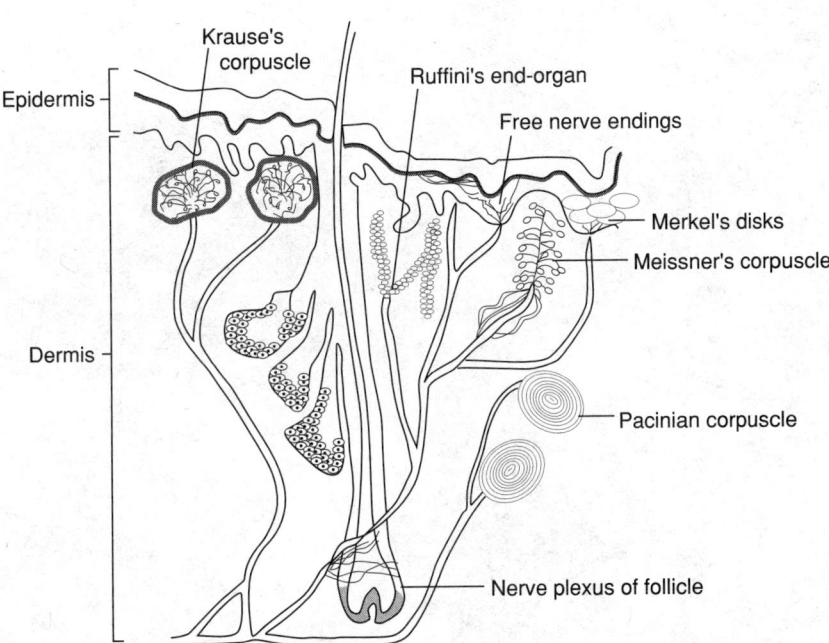

Figure 8-10. Sensory structures of the skin. (Reproduced with permission from Ville CA, Solomon EP, Davis PW: *Biology*. Philadelphia, Saunders College Publishing, 1985, p 880.)

3. **Receptor organs** discriminate between different types of stimuli (Figure 8-10).

   a. **Pacinian corpuscles** respond to pressure changes and are therefore able to detect vibration.

   b. **Meissner's corpuscles** are located primarily in hairless regions of the body (especially the hands and feet) and are sensitive to fine touch, as are **Merkel's disks.**

   c. **Ruffini's end-organs** are found in subcutaneous tissue. They respond to heat and touch.

   d. **Krause's corpuscles** are activated by changes in temperature. For further information on the anatomy of skin, refer to Chapters 7 and 16.

The Endocrine System

I. Function of the Endocrine System

A. Introduction

Like the nervous system, the endocrine system is involved in maintaining homeostasis. Both the nervous system and the endocrine system modulate, integrate, and control the activities of the body. Both systems also enable the body to respond to the environment. However, whereas the effects of the nervous system are rapid and short lived, those of the endocrine system are slower and longer lasting. To control homeostasis, the endocrine system involves feedback loops that are similar to those in the nervous system. Hormones secreted by the endocrine glands regulate organ physiology, and the functioning of the organ systems regulates the secretion of hormones. These feedback loops can be positive or negative, and they are constantly monitored and adjusted.

B. Hormones

Hormones are controlled substances, produced by ductless glands or a collection of cells, which are transported in the circulation to target cells.

1. **Chemical classes**

   a. **Amines** are hormones derived from the amino acid tyrosine. Examples include epinephrine, norepinephrine, and the thyroid hormones triiodothyronine (T_3) and thyroxine (T_4).

   b. **Peptides or glycopeptides** are hormones that are short chains of amino acids. Examples include oxytocin and vasopressin.

   c. **Lipid hormones** are highly hydrophobic, which enables them to easily cross biologic membranes.

   (1) **Prostaglandins** are lipid hormones synthesized from fatty acid precursors. Some of their functions include:

   (a) Decreasing blood pressure

   (b) Inducing labor

   (c) Suppressing gastric secretions

   (d) Mediating inflammation

   (2) **Steroids** are hormones synthesized from a cholesterol precursor. Examples include cortisol and the reproductive hormones estradiol and testosterone.

- d. **Iodinated hormones** are coupled to inorganic iodine. The thyroid hormones T_3 and T_4 are examples.

2. **Mechanisms of action**

   a. **Target specificity** of the endocrine system is a function of specific **receptors** expressed by cells in the body. Hormones released from cells of the endocrine system bind to these receptors, thereby causing a change in the physiology of that cell. This change often includes the activation of specific genes and an induction of protein synthesis.

   b. **Hormone receptors** can be in one of **three locations.**

   (1) The **plasma membrane of the target cell** contains receptors to which amino acid–containing hormones (e.g., amine, peptide, protein classes) bind (Figure 9-1A). This ligand–receptor interaction triggers activation of a second messenger, such as cyclic adenosine monophosphate (cAMP) or guanosine triphosphate (GTP)-binding proteins, which initiates a cascade of events with multiple physiologic outcomes.

   (2) Hormone receptors can also be **inside a cell.** For instance, steroid hormones cross the plasma membrane and bind to receptors in the **cytoplasm of the target cell** (see Figure 9-1B). This ligand–receptor complex then travels into the nucleus and activates gene transcription.

   (3) Receptors can also be present **within the nucleus** directly. Thyroid hormones bind to receptors localized in the nucleus, activating gene transcription (see Figure 9-1B).

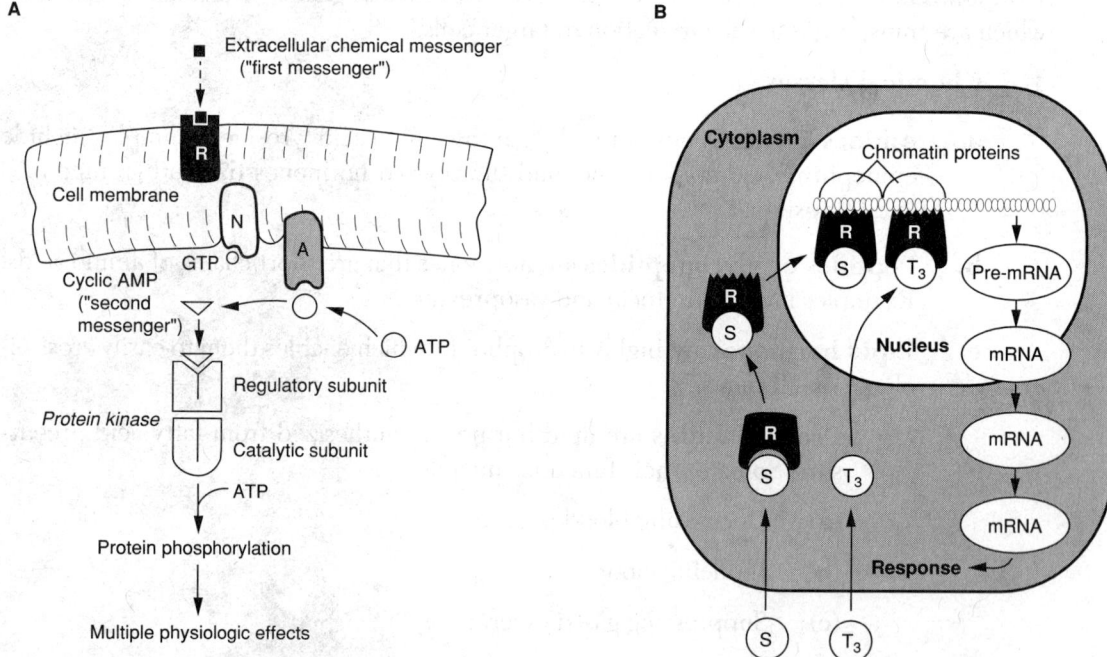

Figure 9-1. Mechanisms of hormone action. **A)** The binding of a hormone to a cell membrane receptor leading to the activation of a protein kinase. **B)** The binding of hormones to cytoplasmic or nuclear receptors and the activation of gene transcription. GTP = guanosine triphosphate; ATP = adenosine triphosphate; A = adenylate cyclase; N = nucleotide regulatory protein; R = receptor; S = steroid hormone; T_3 = thyroid hormone. (Reproduced with permission from Ville CA, Solomon EP, Davis PW: *Biology.* Philadelphia, Saunders College Publishing, 1985, p 906.)

3. **Transport and elimination**
   a. **Many hormones are only active when in a free state.** When hormones are bound to protein, they oftentimes become inactive.
   b. **Free hormones are constantly removed** by target tissues, the liver (which degrades them), and the kidneys (which excrete them). As a result, hormones are **continuously secreted** in small amounts so that enough free hormone is available to activate the necessary target cells (because protein-bound hormone is inactive).

II. Major Endocrine Glands

A. Pituitary Gland

The master gland of the endocrine system, the pituitary gland, is divided into two lobes, anterior and posterior. The functions of the pituitary are closely linked to hypothalamic regulation.

1. The **anterior pituitary** (Figure 9-2A) develops from oral ectoderm and is connected to the hypothalamus by a portal circulatory system, which is a network of blood vessels containing two capillary beds.

   a. The **normal circulatory pathway** involves blood flowing from an artery, through a capillary bed, and into a vein.

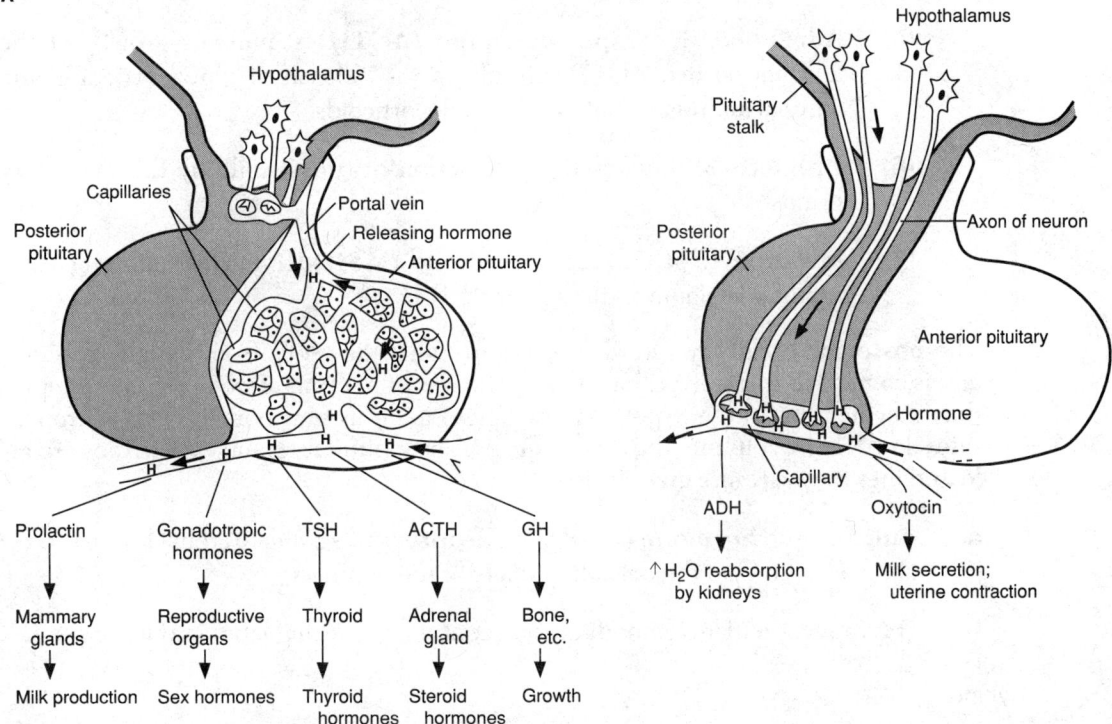

Figure 9-2. The anatomy of the pituitary gland. **A)** The cells of the hypothalamus produce releasing factors that travel through a portal system to reach the anterior pituitary. The anterior pituitary gland synthesizes its hormones after being stimulated by the releasing factors. **B)** The posterior pituitary gland acts as a storage and release site only. Neurons in the hypothalamus produce the posterior pituitary gland hormones and transport them to the posterior pituitary gland by axonal transport. *TSH* = thyroid-stimulating hormone; *ACTH* = adrenocorticotropic hormone; *GH* = growth hormone. (Reproduced with permission from Ville CA, Solomon EP, Davis PW: *Biology*. Philadelphia, Saunders College Publishing, 1985, p 919.)

- **b.** A **portal system** involves blood flowing from an artery, through a capillary bed, into a vein, into a second capillary bed, and into a second vein. The **hypothalamus** secretes peptide-releasing factors into this portal system of blood, which stimulates the cells of the anterior pituitary to secrete the following hormones:

    (1) **Growth hormone (GH)** is a glycoprotein that stimulates growth of bone and muscle and increases blood sugar.

    (2) **Follicle-stimulating hormone (FSH)**

        (a) In **women,** FSH stimulates the growth of follicles and the maturation of oocytes.

        (b) In **men,** FSH stimulates spermatogenesis.

    (3) **Luteinizing hormone (LH)**

        (a) In **women,** LH stimulates ovulation and the formation of the corpus luteum. It also regulates the production of estrogen and progesterone by the follicular cells.

        (b) In **men,** LH stimulates testosterone production by the Leydig cells.

    (4) **Thyroid-stimulating hormone (TSH)** stimulates production of thyroid hormones T_3 and T_4 by the thyroid gland. It also regulates iodination of thyroid hormones.

    (5) **Adrenocorticotropic hormone (ACTH)** stimulates growth of the adrenal cortex. ACTH stimulates the release of glucocorticoids and adrenal androgens but not mineralocorticoids.

    (6) **Prolactin** stimulates the production of breast milk by the mammary glands.

    (7) **Melanocyte-stimulating hormone (MSH)** stimulates melanocytes to produce melanin (skin pigment).

2. The **posterior pituitary gland** (see Figure 9-2B) develops from neural ectoderm and is connected to the hypothalamus by a series of neurons. Two hormones are produced by neurons in the hypothalamus. These neurons transport the two neuropeptides down their axons into the area of the posterior pituitary gland, where they are released into the **systemic circulation.**

    a. **Antidiuretic hormone (ADH),** or **vasopressin,** stimulates renal tubules to reabsorb water from the collecting ducts of the nephron.

    b. **Oxytocin** stimulates smooth muscle contraction during labor and lactation.

B. Thyroid

1. **Structure.** Located on either side of the trachea, the thyroid gland contains **two lobes** connected by an isthmus.

2. **Types.** Two types of hormones are produced by the thyroid gland.

a. **Thyroid hormones** T_3 and T_4 have the following functions:

(1) Control metabolic rate

(2) Regulate growth, differentiation, and maturation

(3) Modulate the nervous system

b. **Calcitonin** causes bone absorption of calcium with a subsequent decrease in serum calcium levels. It also inhibits the activity of osteoclasts, which also lowers serum calcium.

C. **Parathyroid Gland**

1. **Structure.** There are usually **two pairs of parathyroid glands** embedded within the tissue of the thyroid gland.

2. **Parathyroid hormone.** Parathyroid hormone (PTH) is the only hormone made in the parathyroid glands. PTH **increases blood calcium levels** via:

   a. Bone reabsorption by osteoclasts

   b. Decrease in calcium excretion and increase in calcium reabsorption in the kidneys

   c. Activation of vitamin D

D. **Pancreas**

The pancreas has both an **exocrine** and an **endocrine** function. The exocrine activities are discussed in Chapter 12. Endocrine activities are conducted through the **islets of Langerhans,** which secrete the following substances.

1. **Glucagon** is made by alpha cells. It raises the blood sugar level by stimulating glycogenolysis and gluconeogenesis in the liver. Glucagon also mobilizes fat from adipose tissue.

2. **Insulin** is made by beta cells. It lowers the blood sugar level by stimulating glycogenesis in the liver. It facilitates glucose uptake and use of glucose by cells, and it stimulates fat storage and protein synthesis.

3. **Somatostatin** is made by delta cells. It inhibits secretin production by the stomach and pancreas, and it slows gastrointestinal motility. Somatostatin also inhibits the release of GH.

E. **Adrenal Glands**

The adrenal glands are a set of structures resting on the superior border of the kidneys. They are composed of both an outer cortex derived from mesoderm and an inner medulla derived from ectoderm (Figure 9-3).

1. The **adrenal cortex** is subdivided into three layers: **zona glomerulosa, zona fasciculata,** and **zona reticularis.** Each zona produces different hormones.

   a. **Aldosterone** is produced by the **zona glomerulosa.** It is a mineralocorticoid that **regulates ion balance** in the kidney via:

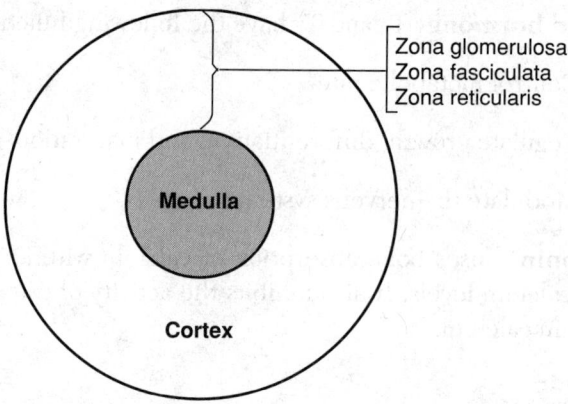

Figure 9-3. Cross section of an adrenal gland. The outer layer of the gland is the cortex. The inner layer is the medulla.

      (1) Stimulation of sodium and chloride reabsorption

      (2) Potassium and phosphorus secretion

      (3) Water retention in the distal convoluted tubule of the nephron

   b. **Cortisol** is produced by the **zona fasciculata.** It is a glucocorticoid "stress" hormone that:

      (1) Increases blood sugar via conversion of amino acids into sugar instead of protein

      (2) Depresses the immune system

      (3) Inhibits the inflammatory response

   c. **Sex steroids (androgens and estrogens)** are produced by the **zona reticularis.** They mediate secondary sex characteristics and regulate reproductive functions.

2. The **adrenal medulla** contains neurons that synapse within the gland and release **epinephrine** and **norepinephrine.** These hormones help the body cope with stress by:

   a. Increasing heart rate and blood pressure

   b. Increasing blood sugar concentration

   c. Increasing metabolic rate

   d. Mobilizing fat stores

The Circulatory System

I. Overview of the Circulatory System Components

A. Function

The circulatory system is a closed system of vessels that serves several functions, including:

1. **Delivery of nutrients and oxygen** to the body tissues
2. **Removal of waste products** from the tissues
3. **Maintenance of body temperature** via thermoregulation (see Chapter 16 of the Biology Review Notes)
4. **Transportation of blood cells**
5. **Delivery of hormones** from their site of production to their target tissues

B. Structures

1. The **heart** is a four-chamber organ that propels blood through the circulatory system (Figure 10-1).

    a. The heart tissue is comprised of **cardiac muscle cells,** which are interconnected by **intercalating disks.**

    b. Blood enters the **atria** of the heart from the venous circulation and leaves the heart in **ventricles** via the arterial circulation.

    c. **Valves** between the atrium–ventricle pair and ventricle–artery pair prevent the backflow of blood during contraction and drive the flow of blood in one direction.

2. The **conduction system** (Figure 10-2) is contained in the heart.

    a. The **sinoatrial (SA) node** is a small group of cells found at the junction of the superior vena cava and the right atrium. The SA node is the **pacemaker of the heart;** it spontaneously depolarizes approximately 100 times every minute. **Depolarization causes atrial contraction.**

    (1) **Parasympathetic innervation** via the vagus nerve slows the heart rate to approximately 70 beats per minute in a resting state.

    (2) **Sympathetic innervation** accelerates the extra rate of depolarization.

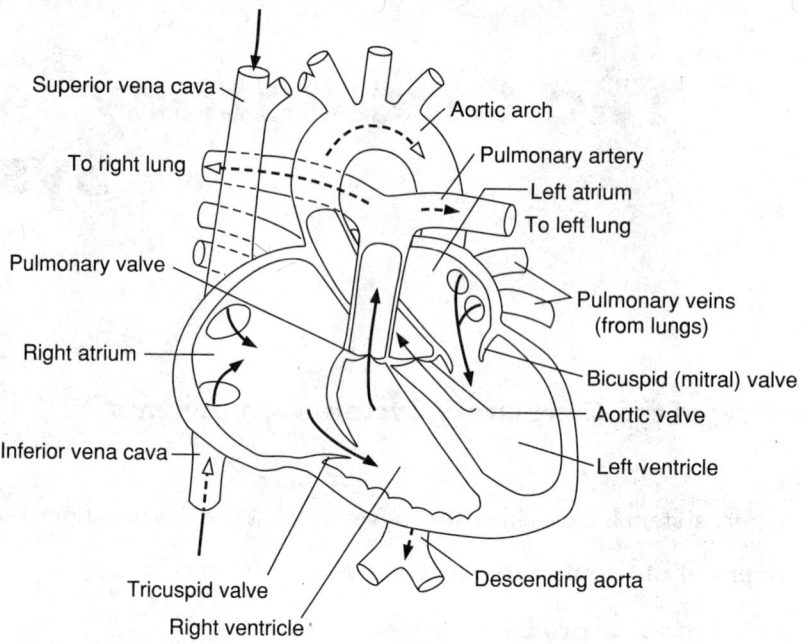

Figure 10-1. Anatomy of the adult heart.

b. The **atrioventricular (AV) node** is at the junction between the right atrium and the right ventricle. Atrial depolarization causes the AV node to depolarize.

c. The **bundle of His** carries the signal from the AV node through the interventricular septum.

d. The **left** and **right bundle branches** carry the impulse from the bundle of His to each ventricle, respectively.

e. **Purkinje fibers** transmit the impulse from the bundle branches into the tissue

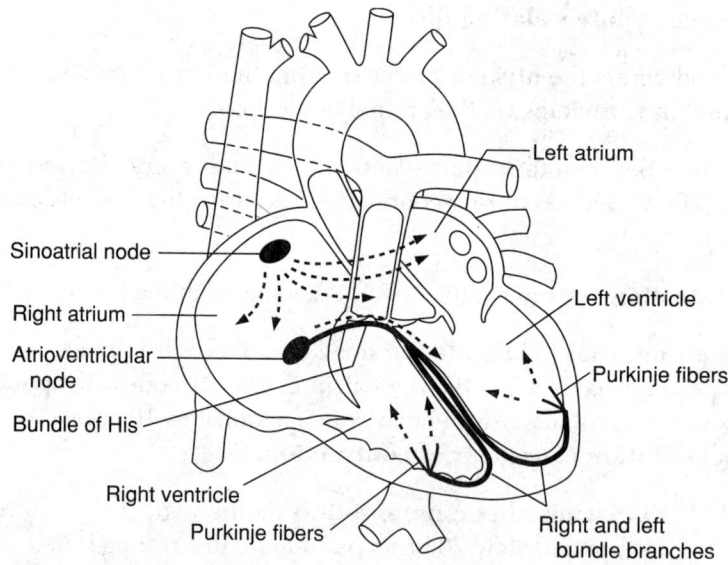

Figure 10-2. Conduction system of the heart. (Reproduced with permission from Ville CA, Solomon EP, Davis PW: *Biology*. Philadelphia, Saunders College Publishing, 1985, p 749.)

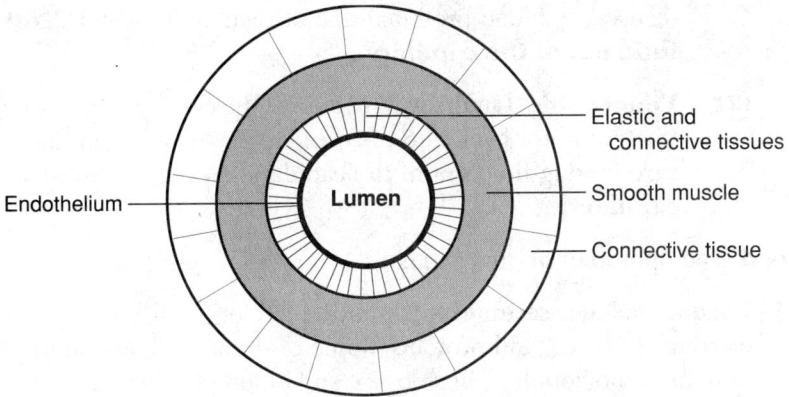

Figure 10-3. Cross section of a blood vessel.

of the ventricles, which causes depolarization and, ultimately, ventricular contraction.

3. **Blood vessels** are composed of several layers of cells (Figure 10-3).

   a. **Arteries** carry oxygenated blood away from the heart (except for the pulmonary artery, which carries deoxygenated blood). Arteries have **higher internal pressures** than veins. Arteries **do not contain valves,** but they do have thick layers of smooth muscle and connective tissue.

   b. **Veins** carry deoxygenated blood toward the heart (except for the pulmonary vein, which transports oxygenated blood). They have **lower internal pressures** than arteries. Veins **contain valves** that help to move blood against the force of gravity. They have a thick elastic layer and thin layers of smooth muscle and connective tissue. Veins **control blood pressure by controlling blood volume.**

   c. **Capillaries** connect arterioles with venules and are the site of exchange between the circulation and the body tissues. **Capillary dynamics** describes the movement of fluid, nutrients, and waste products into and out of the capillary bed (Figure 10-4). More fluid leaves the capillary than returns to it.

      (1) **Arterial side.** Blood entering the arterial side of the capillary has a high hydrostatic pressure that forces fluid out and a low osmotic pressure

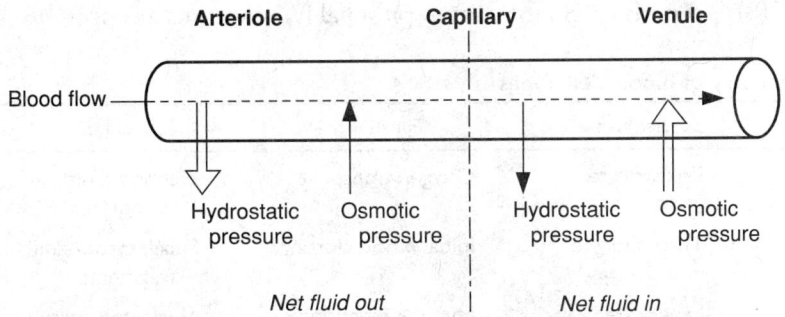

Figure 10-4. Capillary dynamics. Note that the hydrostatic pressure is the dominant force acting on the arterial side of the capillary, whereas osmotic pressure is the dominant force acting on the venous side of the capillary. More net fluid exits than enters the capillary.

(caused by blood proteins) that attracts fluid (water). **Net movement is fluid out of the capillary.**

    (2) **Venous side.** On the venous side of the capillary, the osmotic pressure attracting water back into the capillary is greater than the hydrostatic pressure forcing fluid out of the capillary. **Net movement is fluid into the capillary.**

4. **Blood** is composed of plasma and cells.

   a. **Plasma** includes serum (blood fluid), platelets (cell fragments that assist in the clotting of blood), and proteins. Some of the more abundant proteins are albumin, immunoglobulins, fibrinogen, and blood-clotting enzymes.

   b. Types of **blood cells** are summarized in Table 10-1. White blood cells are discussed in Chapter 11.

5. **Lymphatics** regulate the return of fluid to the circulation.

   a. **Function.** Because more fluid leaves the capillary beds than returns, the body needs a means of returning the excess fluid to the circulatory system. This task is accomplished by the **lymphatic system.**

   b. **Structures**

       (1) The **lymphatics,** or **lymph vessels,** comprise a system of blind-ended channels that carry water, electrolytes, proteins, and waste material back to the bloodstream. Because lymph vessels contain valves but no muscle layer (Figure 10-5), they rely on the skeletal muscle in which they are embedded to propel fluid back toward the heart.

       (2) The **lymph,** or fluid, is filtered by **lymph nodes** and **lymphatic tissues,** which aid the immune system in identifying foreign agents (see Chapter 11).

   c. **Anatomic divisions**

       (1) **Central.** The sites of formation and maturation of white blood cells are the bone marrow and the thymus.

       (2) **Peripheral.** The sites of initial sensitization of immune cells include the lymph nodes, tonsils, Peyer patches, and appendix.

       (3) **Tertiary.** Solitary interepithelial lymphocytes are present.

TABLE 10-1. Summary of Blood Cell Types

Cell Type	Other Name	Function	Appearance	Life Span
Red blood cell	Erythrocyte	Oxygen transport	Biconcave disks; no nucleus	120 days
Platelet	Thrombocyte	Initial blood clotting	Small, cytoplasmic fragments	10 days
White blood cell	Leukocyte	Immune function	Nucleated cells; depend on cell type	Variable

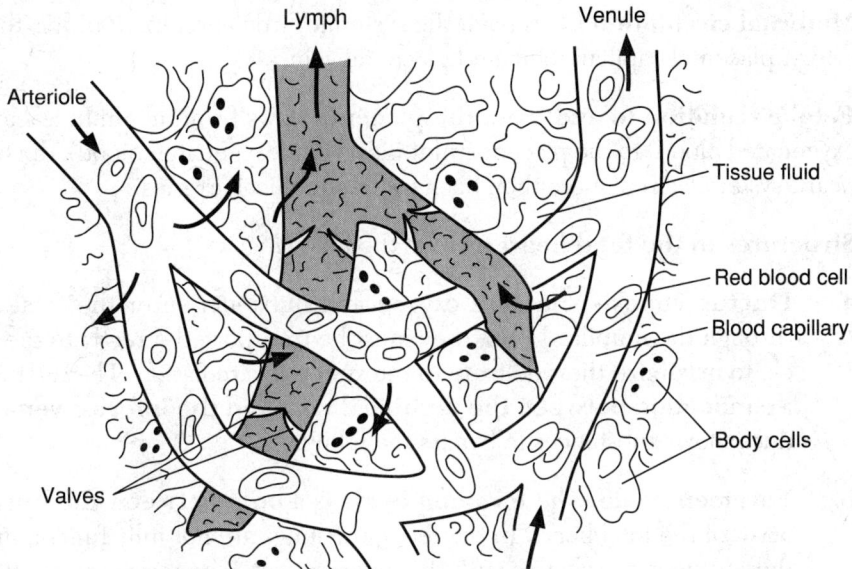

Figure 10-5. Structure and relationship of lymph vessels. (Reproduced with permission from Ville CA, Solomon EP, Davis PW: *Biology*. Philadelphia, Saunders College Publishing, 1985, p 761.)

II. Mechanisms of Circulation

A. Adult Circulation

Blood returns from the venous circulation and empties into the right atrium. Table 10-2 lists the structures that the blood passes through or passes by during circulation.

B. Fetal Heart Circulation

The overall goal of the fetal circulation is to **deliver maternal nutrients and oxygen** to the left side of the fetal heart. Fetal circulation has several structures that are not found in adult circulation. Although blood cells are not exchanged between a mother and her fetus, nutrients, waste products, antibodies, such as immunoglobulin G (IgG), and blood gases are exchanged.

TABLE 10-2. Structures of Adult Blood Circulation*

Cardiac Circulation	Systemic Circulation	Pulmonary Circulation
Right atrium	Heart	Heart
Tricuspid valve	Aorta	Pulmonary artery
Right ventricle	Arteries	Lungs
Pulmonic valve	Capillaries	Pulmonary vein
Pulmonary artery	Veins	
Pulmonic circulation	Vena cavas	
Pulmonic vein		
Left atrium		
Mitral valve		
Left ventricle		
Aortic valve		
Aorta		

*Structures are listed in the order that they are involved in circulation.

1. **Maternal circulation** flows from the systemic circulation to, in order, the placental artery, placental capillary bed, and placental vein.

2. **Fetal circulation to and from the placenta** flows from the umbilical arteries (deoxygenated blood) to the placenta, umbilical vein (oxygenated blood), ductus venosus, heart, systemic circulation, then back to the umbilical arteries.

3. **Structures in the fetal heart** include the following:

    a. **Ductus venosis.** Because oxygen and nutrients enter the fetal circulation through the umbilical vein, they must be transported directly to the left ventricle to maximize their delivery to the systemic circulation. The ductus venosis is a **connection between the umbilical vein and the inferior vena cava.** This duct allows substances to bypass the fetal liver.

    b. **Foramen ovale.** The **foramen ovale** is a **hole between the left and right atria** of the fetal heart. In the fetus, the lungs are not fully functioning because there is no gas exchange with the environment. The resistance of the collapsed lungs to blood flow in the fetus is high. Therefore, bypassing the pulmonary circulation is for delivering the bulk of the nutrients to the systemic circulation and for conserving energy.

    c. **Ductus arteriosus.** Because not all of the blood entering the right heart is transported through the foramen ovale, the ductus arteriosis assists in shunting blood that enters the right ventricle. The ductus arteriosus is a **connection between the pulmonary artery and the aorta.** Again, this allows for direct delivery of nutrients into the systemic circulation by bypassing the fetal lungs.

4. **Fetal circulation in and around the heart** is as follows:

    a. Some blood entering the right atrium moves through the foramen ovale and into the left atrium. This bypasses the right ventricle and lungs. This blood is ultimately pumped into the aorta by the left ventricle.

    b. Some blood entering the right atrium moves into the right ventricle.

    c. Blood that enters the right ventricle is pumped into the pulmonary artery.

    d. From the pulmonary artery, most blood passes through the ductus arteriosus into the aorta.

III. Blood Pressure

The atria are at a lower average pressure than the ventricles because the muscle walls of the atria are thinner and cannot generate as great a force of contraction. The left side of the heart is at a higher average pressure than the right side because the heart walls are thicker. The left side generates a greater force on contraction.

A. **Systolic blood pressure** is the pressure during heart contraction.

B. **Diastolic blood pressure** is the pressure during heart relaxation and filling.

The Immune System

I. Introduction

Many mechanisms enable the human body to respond to foreign substances. However, the same mechanisms that allow the body to resist bacterial infections, for example, can also cause tissue injury and disease under certain circumstances (e.g., autoimmunity). Therefore, **immunity** is defined as **a reaction to foreign agents** without implying a physiologic or pathologic outcome of this reaction. According to this definition, **foreign agents include microbes** (e.g., bacteria, viruses, fungi, parasites), **microbe-infected host cells, cancer cells,** and **macromolecules** (e.g., protein, polysaccharides).

II. Types of Immunity

A. **Natural immunity** (also called innate or native immunity) describes those defense mechanisms that are present before exposure to foreign agents. Natural immunity reactions are nonspecific, and defense mechanisms include the following:

1. **Physicochemical barriers**—skin and mucous membranes

2. **Molecules in the blood circulation**—complement

3. **Immune cells**—phagocytic cells (macrophages, neutrophils) and natural killer cells

4. **Soluble mediators**—cytokines (regulating substances) derived from immune cells

B. **Acquired immunity** is the stimulation or induction of other defense mechanisms by exposure to foreign substances (i.e., antigens). These reactions are specific for distinct molecules and increase in magnitude and effectiveness with each subsequent exposure to the eliciting stimulus.

1. The acquired, or specific, immune system has a **memory function** in which **reactive cells linger** after the stimulus has been eradicated so they can quickly respond to future encounters. This type of immunity also has the capacity to **amplify natural immunity protective mechanisms** and to **direct these mechanisms to the site of entry** of the antigen.

2. Features of specific immunity include:

    a. **Physicochemical barriers**—cutaneous and mucosal immune systems; antibodies in mucosal secretions

- b. **Molecules in the blood circulation**—antibodies
- c. **Immune cells**—lymphocytes
- d. **Soluble mediators**—lymphocyte-derived cytokines

3. **Types of specific immunity** include:
    a. **Active immunity**—an individual's specific immune reaction on exposure to antigen. A **lag period** is required for the immune system to become activated.
    b. **Passive immunity**—specific immunity conferred on an individual via the transfer of cells or serum from another immune person. Upon transfer, the immune responses are **immediately active.**

4. **Classes of specific immunity** include:
    a. **Humoral immunity**—passive immunity acquired via the transfer of plasma or serum. It is mediated by antibodies released by B cells.
    b. **Cell-mediated immunity**—passive immunity acquired via the transfer of lymphocytic cells.

III. Cells

A. **Lymphocytes** are the basic units of the immune system. They are derived from a **common precursor cell in the bone marrow.**

1. **T cells** are those lymphocytes that mature only after passage through the thymus. They have receptors on their cell surface that enable them to recognize antigens presented by major histocompatibility complexes. T cells regulate cell-mediated immunity. There are several important types of T cells.

    a. **Cytotoxic T cells** cause direct killing of cells that express foreign antigens on their cell surface. As a result, cytotoxic T cells play an important role in the elimination of viral-infected cells, cancer cells, and cells with an intracellular pathogen. Cytotoxic T cells have a $CD8^+$ surface marker.

    b. **Helper T cells** promote the activities of other cells through the **synthesis and secretion of cytokines,** which are soluble factors that bind to receptors on the surface of a target cell and initiate a physiologic change in that cell. Helper T cells have a $CD4^+$ surface marker.

    c. **Suppressor T cells** inhibit the activities of certain cells through cytokine mediators. Suppressor T cells have a $CD4^+$ surface marker.

2. **B cells** are involved in **humoral immunity,** which is a defense mechanism mediated by antibodies. **Antibodies** (Figure 11-1) are produced by mature B cells, known as **plasma cells,** in response to foreign antigens. Antibodies are made by a process of genetic recombination. A specific antibody is generated against each antigen.

    a. The **structure of antibodies** differs based on how many units the antibodies contain. The basic unit is shown in Figure 11-1. Immunoglobulin G (IgG) is a monomer and contains one basic unit. Immunoglobulin M (IgM) is a pentamer, and is made of five monomeric units linked together.

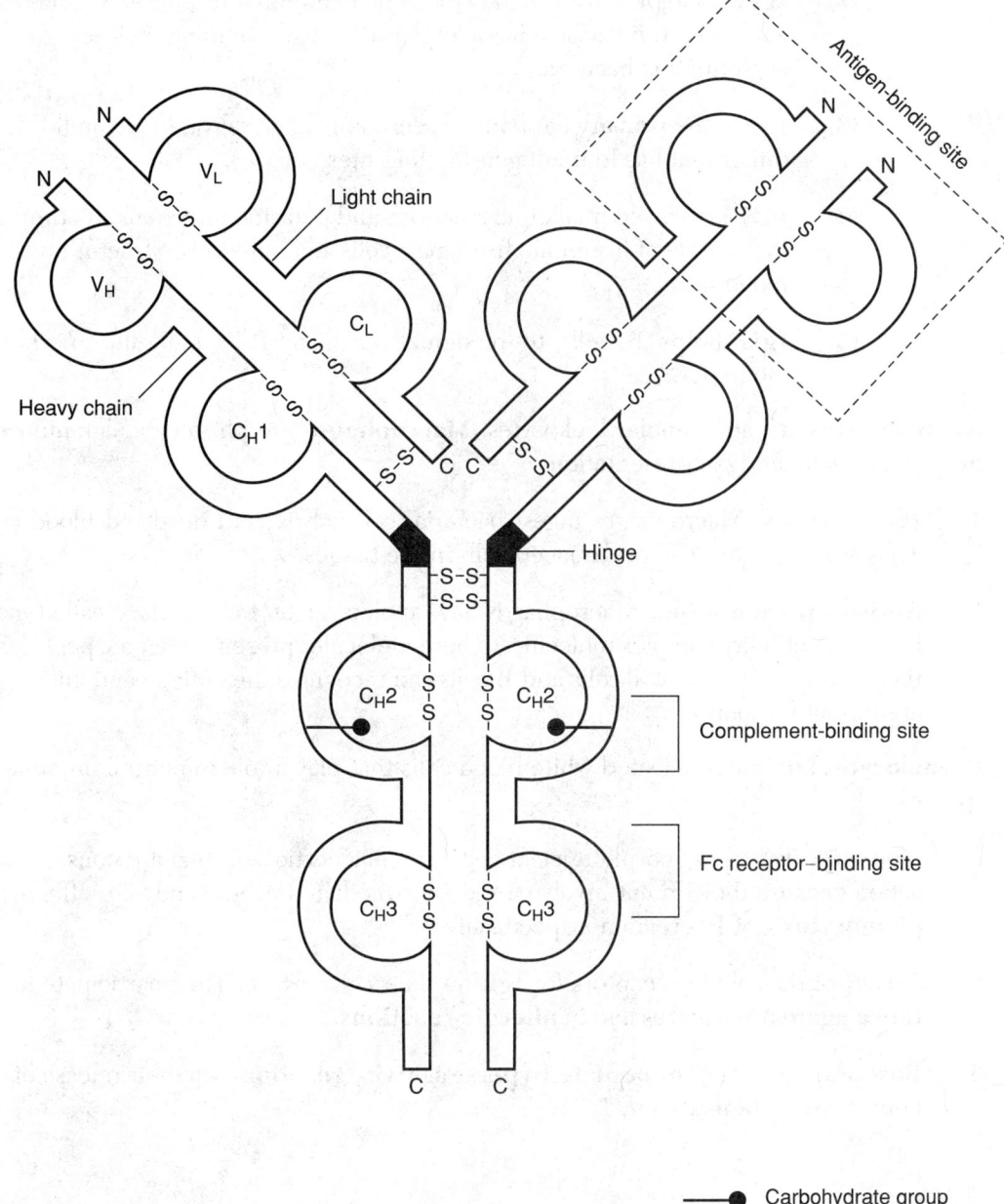

Figure 11-1. Basic antibody structure. The structure shown is immunoglobulin G (IgG), which is a monomer. V = variable region. C = constant region

      b. There are **five classes of antibodies,** with each class containing specific antibodies. Each class of antibody has a somewhat different function. The classes are immunoglobulins (Ig) G, M, A, D, and E.

          (1) **IgG** is the major antibody in the bloodstream. It is usually the second antibody produced in an initial immune response. It crosses the placenta and, as a monomer, it has two antigen-binding sites.

(2) **IgM** is the primary antibody made in an immune response. As stated, it is a pentamer. Because it has a total of 10 antigen-binding sites, it is good for agglutinating bacteria.

(3) **IgA** is the primary antibody in secretions (e.g., saliva, breast milk). It is a dimer that has four antigen-binding sites.

(4) **IgE** is involved in allergic reactions and parasitic infections. It stimulates the release of histamine from mast cells and basophils. Structurally, it is a monomer.

(5) **IgD** helps B cells to recognize antigen. It is a membrane-bound monomer.

B. **Agranulocytes** are nongranular leukocytes. **Macrophages** are phagocytic agranulocytes that play a role in antigen presentation.

1. **Phagocytosis.** Macrophages ingest bacteria, cell debris, and dead red blood cells. They scavenge for dead or damaged cells in the tissues.

2. **Antigen presentation.** Macrophages have molecules on their surface called major histocompatibility complex molecules. These molecules present antigenic peptides on their cell surface so that T cells and B cells can recognize the antigen and initiate the appropriate response.

C. **Granulocytes** are multinucleated white blood cells that play a role in natural immune responses.

1. **Neutrophils,** or polymorphonuclear cells, are phagocytic cells that are nonspecific in action because they do not involve a receptor-mediated event. They are effective in **phagocytosis of bacteria** nonspecifically.

2. **Eosinophils** contain receptors for IgE on their cell surface. They participate in **defense against parasites** and in **allergic reactions.**

3. **Basophils** generate **immediate hypersensitivity reactions** via their release of histamine when stimulated by IgE.

IV. Tissues

A. **Bone Marrow**

Blood cells, including those of the circulatory system, are produced in bone marrow. Some cells (e.g., B cells) complete maturation in the bone marrow; others require passage through other organs before gaining complete function (e.g., T cells must pass through the thymus). The cells of the bone marrow play a role in selecting the immune cells that react against self and subsequently destroying them.

B. **Spleen and Lymph Nodes** (Figure 11-2)

Both the spleen and the lymph nodes are part of the lymphatic system (see Chapter 10).

1. **Lymph nodes.** White blood cells segregate into specialized locations in the lymph nodes. As the lymph is filtered on its return to the circulatory system, foreign antigens

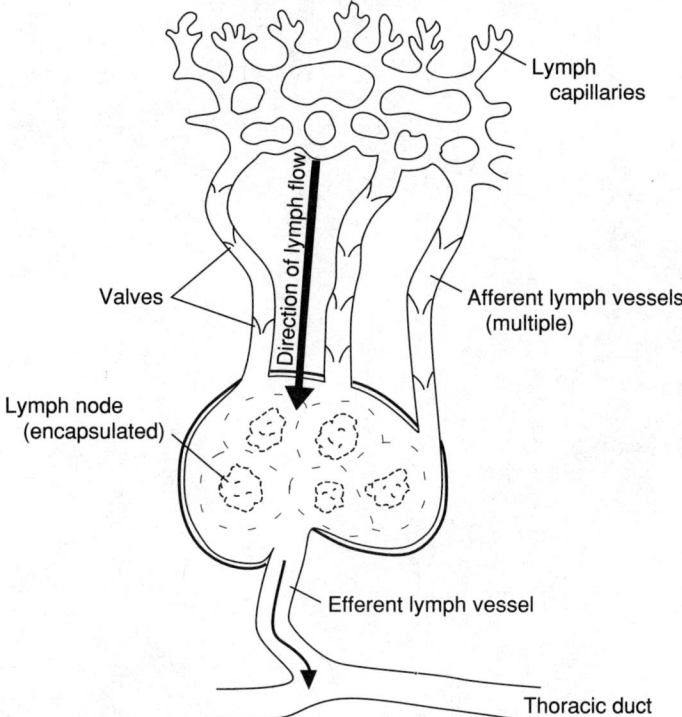

Figure 11-2. Lymph node anatomy. Note that lymph ultimately returns to the bloodstream via the thoracic duct.

are recognized by these immune cells, and an immune response is generated. This keeps bacteria from gaining access to the blood.

2. **Spleen.** The spleen removes debris from circulating blood and clears defective red blood cells from the circulation. It also produces an immune response against foreign antigens.

The Digestive System   12

I. Nutrition

A. Introduction

To meet energy requirements, organisms must ingest materials from the environment; these materials are usually present in the diet as polymers of simpler compounds. For the human body to use dietary components efficiently, it must first break down these complex polymers into their unit constituents. Larger substances are less readily absorbed by the intestinal epithelial cells. Furthermore, small subunits can be used more readily by the body to meet its own synthetic demands. The unit of measure for dietary energy is the **kilocalorie (kcal)**.

B. Nutrients

1. **Carbohydrates** yield 4 kcal of energy per gram and generally provide the greatest source of energy in the diet.

   a. Carbohydrates, or polysaccharides, are long chains of five- or six-carbon sugars called monosaccharides, which are the smallest functional units and cannot be further broken down.

      (1) **Monosaccharides** include glucose and fructose.

      (2) **Polysaccharides** include starch and glycogen.

   b. **Carbohydrate digestion** begins in the oral cavity with **salivary amylase** and is completed in the small intestine by exposure to **pancreatic amylase.** Intestinal epithelial cells absorb monosaccharides and disaccharides, which are further hydrolyzed into their monosaccharide subunits.

2. **Fats** are lipids composed of the three-carbon molecule **glycerol** attached via phosphate linkages to **three fatty acids,** which contain long carbon polymer chains. When these carbon bonds are hydrolyzed, 9 kcal of energy are released per gram.

   a. **Saturated fatty acids** do not contain double bonds, whereas unsaturated molecules have either *cis* or *trans* double bonds.

   b. **Fat digestion** occurs when the hormone **cholecystokinin (CCK),** made in the lining of the duodenum, causes the release of **lipase** from the pancreas and **bile** from the gallbladder. These substances help emulsify the fat and form **micelles,** which are absorbed in the duodenum and jejunum.

3. **Proteins** are composed of chains of amino acids, which are the functional units. During proteolysis, 4 kcal of energy per gram are released from the hydrolysis of the peptide bonds.

   a. **Essential amino acids** must be consumed as part of the diet, whereas some other amino acids can be synthesized by the body.

   b. **Protein digestion** begins in the stomach by the action of **pepsin.** When the protein is transferred to the small intestine, CCK secretion is stimulated. This induces the release of **pancreatic proteolytic enzymes** (i.e., trypsin, chymotrypsin, elastase, numerous types of endopeptidase and exopeptidase). These enzymes complete protein digestion.

4. **Vitamins and minerals** are supplied to the body by the diet.

   a. **Vitamins** are organic molecules that aid in enzyme catalysis by acting as **coenzymes** [e.g., niacin is the coenzyme of nicotinamide adenine dinucleotide (NAD)]. They are either ingested or produced by bacteria within the gastrointestinal tract (e.g., vitamin K, vitamin B_{12}).

   b. **Minerals** are inorganic ions that can also aid in catalysis by acting as **prosthetic groups** (e.g., iron in the heme group of hemoglobin), in addition to playing roles in functions such as protein folding and electrochemical gradients.

II. Gastrointestinal Tract

A. **Passage of Food**

   1. **Muscular control.** Food enters the body via the mouth and is eventually defecated through the anus (Figure 12-1).

      a. **Peristalsis,** which is a rhythmic, coordinated, wave-like movement of involuntary muscle contractions, propels ingested material unidirectionally through the digestive system.

      b. **Distention of a hollow organ** stimulates circular muscles to contract immediately behind the enlarged area. At the same time, the circular muscle in front of the enlarged area relaxes and the longitudinal muscle in between contracts, pushing the food bolus forward.

   2. **Sphincters.** Sphincters are specialized rings of muscle that surround an orifice. The gastrointestinal tract is divided into compartments by several sphincters. Through contraction and relaxation, sphincters control the transition of food from one area to another.

      a. The **pyloric sphincter** governs the rate at which food enters the small intestine from the stomach.

      b. The **esophageal sphincters** prevent acidic stomach contents from moving back up into the esophagus.

      c. When there is **mechanical failure,** that is, when the esophageal or pyloric sphincters function abnormally, stomach acid can leak into the esophagus (heartburn) or the duodenum (ulcer formation).

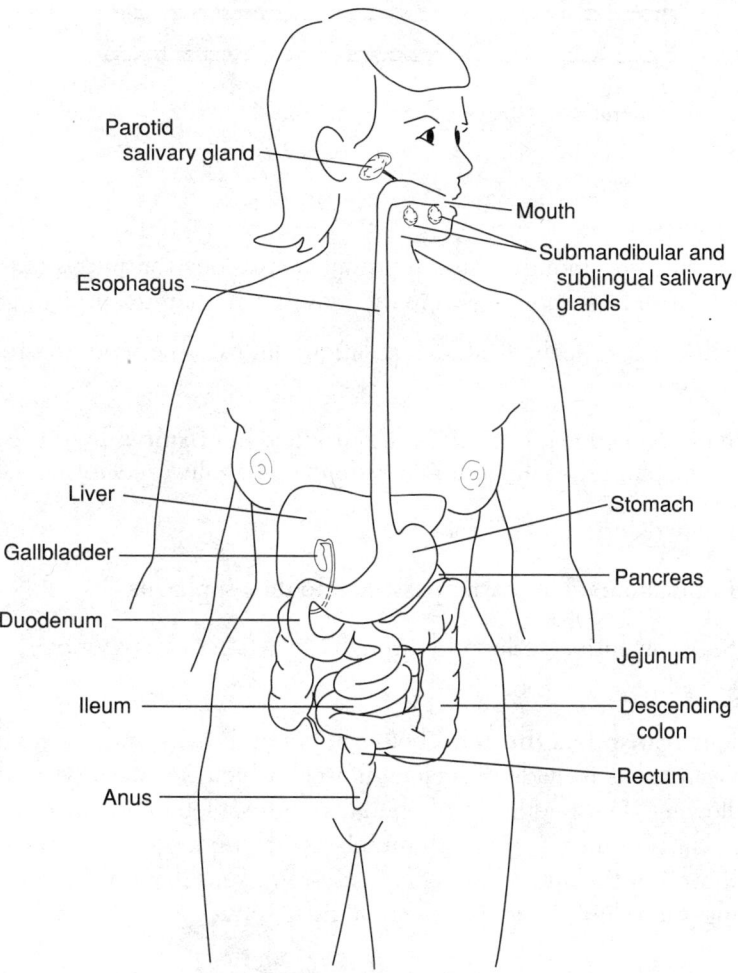

Figure 12-1. The passage of food through the gut. A diagram of some of the major structures of the digestive tract.

3. **Secretions.** Enzymes assist in the digestion of nutrients. To make contact with the food, enzymes travel through ducts that empty into the gut lumen.

   a. **Digestive enzymes are proteins.**

      (1) Many digestive enzymes are released as **inactive precursors** called **zymogens,** or **preproenzymes** (Figure 12-2), which become active only when they reach the lumen of the gut.

      (2) **Secretion of an inactive form** is thereby a means of **protection against self-digestion.**

      (3) For example, **trypsin** is activated in the gut lumen only by stomach acid or by self-proteolysis by another trypsin molecule. Activated trypsin then goes on to activate the other digestive enzymes.

   b. **The stimuli of gastrointestinal enzymes**

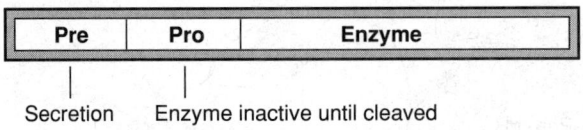

Figure 12-2. The structure of zymogens.

B. **Mouth and Pharynx**

Food is taken into the mouth where it stimulates various sensations (e.g., taste, texture, odor, temperature). Digestion begins in the mouth and pharynx through two means.

1. **Mastication** is a mechanical manipulation that uses the teeth to break the food into smaller pieces.

2. Enzymatic breakdown occurs via the production of **saliva** by three pairs of salivary glands (i.e., parotid, submandibular, sublingual). Saliva acts to:

    a. Begin digestion of carbohydrates

    b. Lubricate food for easier passage into the esophagus

    c. Serve an antibacterial role through the action of **lysozymes**

C. **Esophagus**

The esophagus **transports the food** between the oral cavity and the stomach. The upper third of the esophagus is made of skeletal muscle, which is necessary for the voluntary action of swallowing. The middle third contains both skeletal and smooth muscle, and the lower third is smooth muscle. Smooth muscle, which undergoes involuntary movement, is under the control of the autonomic nervous system (see Chapter 8). Smooth muscle lines the remaining gut wall of the gastrointestinal tract.

D. **Stomach**

The stomach stores, churns, and digests food. Digestion is accomplished through the **release of enzymes and acid.**

1. **Gastrin** is a hormone that is released early in the digestive process.

    a. **The release of gastrin from G cells is induced by:**

        (1) Stomach distention

        (2) Parasympathetic stimulation

        (3) Proteins in the stomach lumen

    b. **Gastrin stimulates:**

        (1) Stomach distention and motility (increased smooth muscle contraction and peristalsis)

        (2) The release of **pepsinogen,** which is converted to active **pepsin** by hydrochloric acid (HCl) and by pepsin proteins, from the chief cells

        (3) The secretion of HCl from parietal cells

2. To prevent damage to the stomach lining by acid, both **bicarbonate** (HCO_3^-) and **mucus** are produced, which, respectively, serve to neutralize the acid and to shield the stomach mucosa.

3. **Intrinsic factor** is a protein that binds to vitamin B_{12} and protects it from acid breakdown until it can be absorbed in the small intestine. Intrinsic factor is secreted by the parietal cells.

E. **Small Intestine**

Digestion is completed in the small intestine, and most of the absorption of nutrients occurs in the jejunum and the ileum.

1. To facilitate absorption, the lining of the intestine has numerous projections of gut mucosa (i.e., **villi**) and microscopic projections from the columnar epithelial cells (i.e., **microvilli**). Both of these modifications increase the surface area of the small intestine, assisting in both the absorption of nutrients and the secretion of mucus and enzymes.

2. The **duodenum** is the first region of the small intestine. It makes up approximately the first 12 inches of the human small intestine.

   a. **Secretin** is a hormone whose release in the duodenum is stimulated by stomach acid. This hormone is produced by cells in the lining of the duodenum. Secretin has the following effects:

      (1) Causes the pancreas to produce and secrete HCO_3^-

      (2) Slows gastric motility and emptying

      (3) Inhibits gastrin release

   b. **CCK** is released in response to fatty acids and protein.

      (1) **CCK stimulates the pancreas to release pancreatic enzymes:**

         (a) **Trypsin** is released for protein digestion and activation of proenzymes.

         (b) **Lipase** is released for fat digestion.

         (c) **Amylase** is released for carbohydrate digestion.

      (2) **CCK stimulates gallbladder contraction and emptying, which releases bile.** Bile is secreted by the liver, stored in the gallbladder, and released into the duodenum. Bile contains bile salts, cholesterol, and bilirubin (which is a product of hemoglobin degradation). Bile salts act like detergent to emulsify fats.

3. The **jejunum** is the upper two thirds of the small intestine. It releases enzymes (e.g., maltase, carboxypeptidase) and absorbs the majority of digesting nutrients.

4. The **ileum** is the lower one third of the small intestine. It completes nutrient absorption, and it reabsorbs bile acids for recycling.

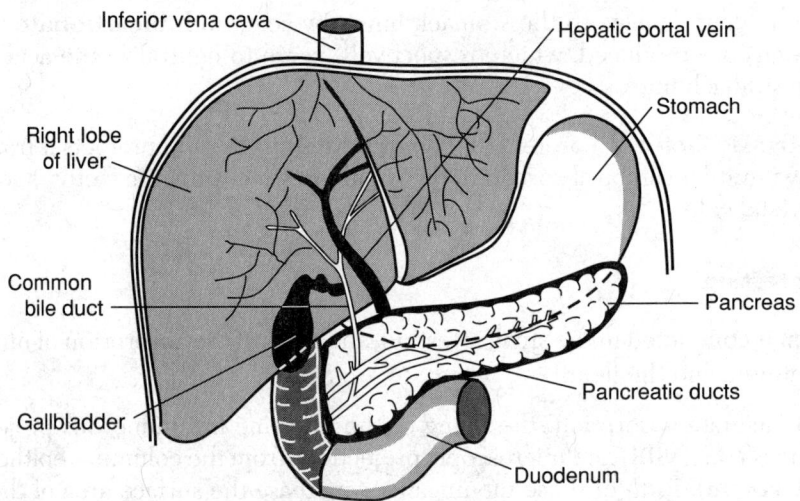

Figure 12-3. Anatomy of the upper gastrointestinal tract. (Reproduced with permission from Ville CA, Solomon EP, Davis PW: *Biology*. Philadelphia, Saunders College Publishing, 1985, p 703.)

F. Colon

The colon has several subdivisions, including the **cecum**, the **ascending colon**, the **transverse colon**, the **descending colon**, and the **sigmoid colon**. The colon serves several functions.

1. The colon **reabsorbs water and electrolytes** and thus has a role in osmoregulation.

2. The colon **forms and stores stool** for elimination from the body.

III. Accessory Organs (Figure 12-3)

A. Liver

The liver is the most important metabolic organ. Functions of the liver are listed in Table 12-1.

TABLE 12-1. Functions of the Liver

Regulation of blood glucose via glycogen storage
Storage of essential vitamins
Interconversion of nutrients
Detoxification of drugs, including alcohol
Protective immunity via Kupffer cells (liver macrophages)
Destruction of RBCs (in addition to spleen)
Production and secretion of bile
Production of urea from ammonia (toxic), a byproduct of protein metabolism
Synthesis of plasma proteins (e.g., albumin, clotting proteins)
Manufacture of blood cells during embryogenesis
Release of HCO_3^-
Manufacture of lipoproteins for cholesterol transport

RBCs = red blood cells.

B. Pancreas

The pancreas serves both an endocrine function (see Chapter 9) and an exocrine function. The exocrine pancreas is involved in digestion through its **production of pancreatic juice.** Pancreatic juice neutralizes stomach acid with bicarbonate ion, and it provides the enzymes necessary for the digestion of carbohydrates, fats, and proteins.

The Excretory System

I. The Role of the Excretory System in Body Homeostasis

The ability to survive in osmotically unstable environments has been achieved by evolution of effective excretory systems. The simplest and smallest aquatic organisms use diffusion into the surrounding water to excrete metabolic wastes. Higher order animals, which have circulatory systems, have evolved kidneys through which blood passes and is filtered. **Kidney functions include:**

1. **Fluid and Electrolyte Balance**
2. **Control of Blood Pressure**
3. **Acid–base Balance**
4. **Stimulation of the Bone Marrow to Produce Red Blood Cells**

II. Mammalian Kidney Structure and Function

A. Gross Anatomy (Figure 13-1)

1. **Location.** The kidneys are paired organs lying behind the abdominal organs on each side of the body. They lie behind the peritoneum, against the ribs of the middle to lower back.

2. **Size.** The human kidney is approximately the size of a fist and accounts for less than 1% of total body weight.

3. **Blood flow and filtration.** The kidneys receive a massive blood flow—approximately 20% of the cardiac output. The kidneys filter plasma up to 125 ml/min, which approaches 170 L/day.

4. **Outer covering.** The kidney is covered with the renal capsule, which is a thin layer of connective tissue.

5. **Two functional regions produce urine:**

    a. The **cortex** is the outer functional layer.

    b. The **medulla** is the inner functional layer.

6. **Associated structures.** A **papilla** extends from the cortex into the medulla. The papilla funnels urine into the **renal pyramids,** which carry urine to the **renal pelvis,**

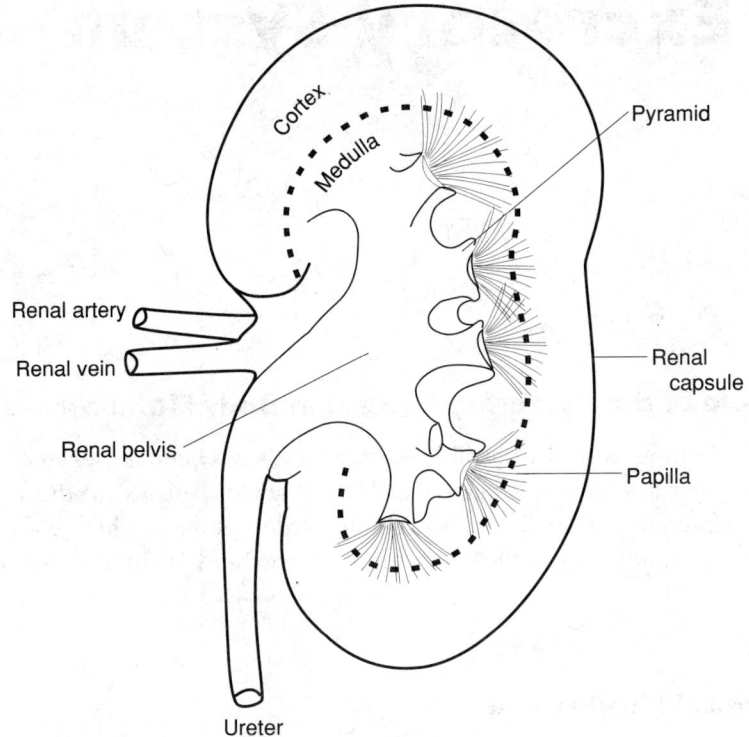

Figure 13-1. The gross anatomy of the human kidney.

which is the hollow collecting chamber of the kidney. The renal pelvis gives rise to the **ureters,** which are tubes that carry urine to the **urinary bladder.** Urine leaves the bladder via the urethra.

B. Microscopic Anatomy and Function

1. The **nephron is the functional unit of the kidney.** The human kidney contains more than 1,000,000 nephrons. The main purpose of the nephron is to allow filtering of blood plasma, allowing resorption and secretion of key electrolytes and nutrients. The end product of this activity is the production of a concentrated urine.

2. To understand the function of the kidney, the **anatomy of the microcirculation** in the kidney must be known (Figure 13-2).

    a. **Toward the kidneys.** Blood enters the kidney from the **aorta** by the **renal artery.** The renal artery branches many times and gives rise to almost 1,000,000 end branches known as **afferent arterioles.** The afferent arterioles lead to a complex capillary network called the **glomerulus.**

    b. **Away from the kidneys.** The **glomerular capillaries** are shaped like a ball of yarn, and they have pores that allow electrolytes and small molecules (i.e., molecules with molecular weights less than 200) to filter out of the blood. The blood that does not filter out of the glomerulus travels through the **efferent arterioles** to the **vasa recta vessels,** which reabsorb water and solutes from the interstitial space. Blood eventually returns to the **renal vein** and the **vena cava.**

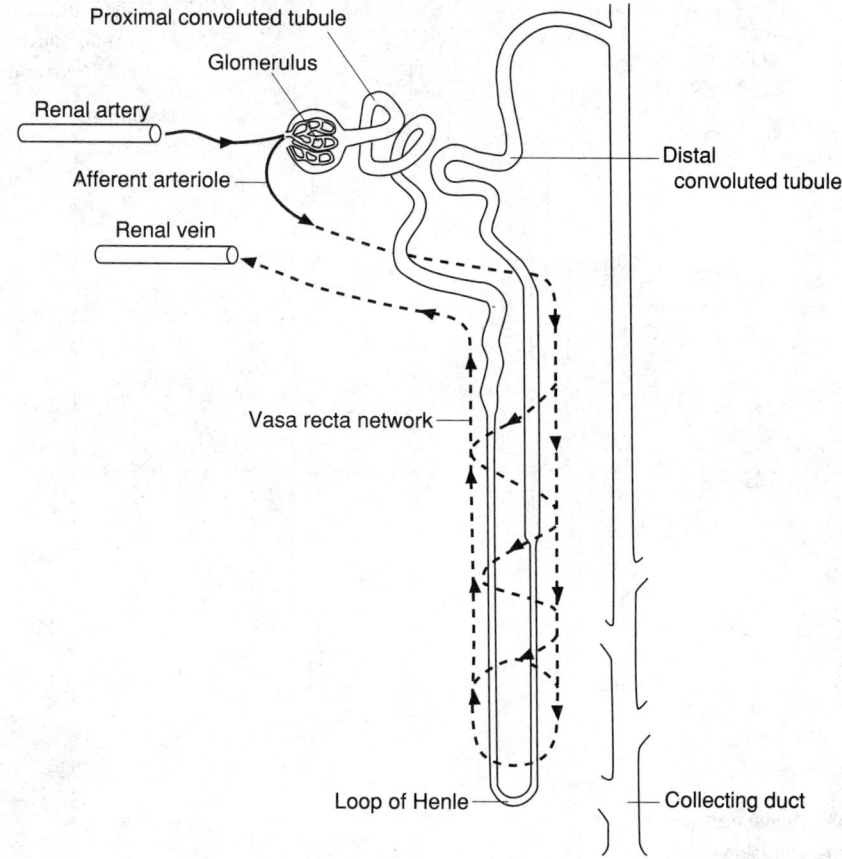

Figure 13-2. The microscopic anatomy of the nephron. Note that the arrows show the direction of blood flow.

3. The **anatomy of the tubules** is also important in nephron function.

   a. The nephron begins at the **Bowman capsule,** which is a structure composed of a single cell layer that is adjacent to the glomerulus. **Blood filtrate,** which filters out of the glomerulus, accumulates in the Bowman capsule.

   b. From the capsule, the filtrate moves into the **proximal convoluted tubule.**

   c. From this tubule, the filtrate travels around a hairpin turn known as the **loop of Henle.**

   d. Filtrate then moves up a **thin ascending limb,** followed by a **thick ascending limb,** eventually reaching the **distal convoluted tubule.**

   e. The distal convoluted tubule empties directly into a **collecting duct,** which eventually drains into the **renal pelvis.**

C. **Formation of Urine** (Figure 13-3)

   1. Urine formation begins with the **ultrafiltration** of blood plasma and the accumulation of ultrafiltrate in the lumen of the Bowman capsule.

      a. **Extensive filtration.** Up to **25% of the water and solutes** flowing through the glomerulus is filtered into the Bowman space.

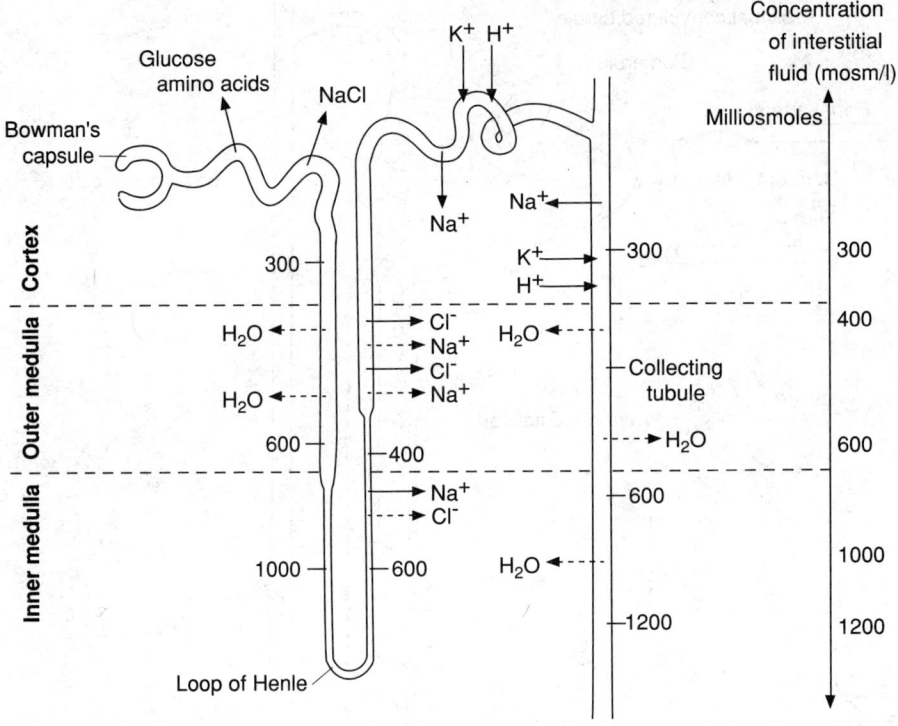

Figure 13-3. The formation of urine.

- **b. Pressure difference.** The driving force for filtration is the difference between hydrostatic pressure (blood pressure) and osmotic pressure in the glomerulus. The difference approaches **45 mm Hg**.
- **c. Weight limitation.** Because of the pore size of the glomerular capillaries, only particles with molecular weights less than several hundred can be filtered. Glucose, electrolytes, amino acids, and urea are filtered, whereas albumin, cells, and other large proteins are not filtered.

2. Of the 170 L of plasma that are filtered per day, the **final urine volume** is only approximately 1.5 L. Thus, more than 99% of water and a large proportion of the electrolytes are reabsorbed.

3. Some substances appear in the urine in a concentration higher than in the blood. These substances are **selectively secreted** from the renal tubules into the urine.

4. From the Bowman capsule, the filtrate moves into the **descending limb of the loop of Henle.**
   - a. The descending limb has very low permeability to sodium and chloride ions, and no active transport occurs there.
   - b. This tubule is very permeable to water.
   - c. A strong NaCl gradient exists in the interstitial space in this region.

5. The **thin ascending limb** is permeable to sodium and chloride ions, and some investigators believe that active transport of sodium ions occurs there.

6. The **thick ascending limb** shows almost no permeability to water and exhibits active transport of chloride ions from the lumen of the tubule into the interstitial space. Sodium ions passively follow.

7. The **distal convoluted tubule** is very active in transport.

    a. **Active transport of sodium ions** followed by the **passive movement of chloride ions** occurs there.

    b. Other distinct active transport proteins allow for a variety of other ions to be transported.

    (1) **Potassium–hydrogen ion cotransport** occurs. Potassium ions are excreted into the lumen from the tubule cell along with hydrogen ions. This cotransport provides a mechanism by which the body can excrete potassium ions and control its acid–base balance with hydrogen ions.

    (2) The steroid hormone **aldosterone** plays a role in controlling the sodium ion and potassium–hydrogen ion transport in the distal tubule and in the collecting duct.

8. The **collecting duct** removes water from the hypotonic fluid entering the distal tubule and thereby produces hyperosmotic urine.

    a. **Antidiuretic hormone (ADH), or vasopressin,** must be present for the collecting duct to be permeable to water.

    b. In the **absence of ADH,** the duct is impermeable to water, and much water from the contents of the lumen is lost.

9. **Urine that leaves the collecting duct is concentrated** and contains the following: large amounts of nitrogenous waste (i.e., urea), less than 1% of the water and salts that were filtered, and secreted substances (e.g., drugs, toxins).

D. **Concentration Mechanism**

Although microscopically thin, the nephron is very long and spans from the cortex to the medulla of the kidney. Because the descending limb is water permeable but salts do not leave the tubule lumen, the osmolarity of urine increases as urine travels down the descending limb and is maximum at the loop of Henle. The loop of Henle is water impermeable, and salts stay in the tubule lumen.

1. As salts either leave the tubule passively or actively in the water-impermeable ascending limb, the urine osmolarity decreases. Finally, when urine passes down the collecting duct, it becomes **reconcentrated** because a tremendous amount of water leaves the urine (via ADH), concentrating the remaining solutes.

2. A **countercurrent feature** in the organization of the circulation of the nephron aids in maintaining the concentration gradient of the interstitium.

    a. The flow of urine in the tubules is opposite the flow of blood in the vasa recta (see Figure 13-2).

        b. Vasa recta blood flow begins at the thick ascending limb and flows around the loop of Henle to the thin descending limb.

        c. By selective resorption of salts and water, this countercurrent system aids in maintaining the concentration gradient of the interstitial fluid.

   3. The **vasa recta vessels** form loop-like networks around each nephron.

        a. **Blood descends from the cortex** into the deeper portions of the medulla and forms the vasa recta networks.

        b. The **vasa recta then ascends toward the cortex.** Moving from the cortex to the medulla, blood in the vasa recta takes up salt and gives up water osmotically to balance the increasing osmotic load of the interstitial fluid.

        c. **As blood returns toward the cortex** from the medulla in the vasa recta network, the blood gives up salt and takes in water as it encounters an interstitial fluid of progressively lower osmolarity.

        d. The purpose of this system is to **maintain the concentration gradient of the interstitial fluid.** The vasa recta takes up salts and gives up water in the salty medulla, and it gives up salt and takes in water in the dilute cortex.

E. Hormonal Influences on the Nephron

   1. **Aldosterone,** which is a steroid hormone produced in the adrenal cortex, acts on the distal convoluted tubule of the nephron.

        a. Aldosterone increases the active transport of **potassium ions** into the urine for **excretion** and of **sodium ions** from the urine into the cell for **reabsorption.**

        b. Aldosterone plays a major role in the **regulation of blood pressure** by controlling the level of sodium in the blood. Sodium retention in the body increases blood pressure by holding body water.

   2. **ADH allows water reabsorption** from the collecting duct of the nephron.

        a. ADH is believed to **create pores** in the cells of the duct to allow massive water permeability and reabsorption.

        b. ADH is controlled by a **feedback regulation system.** Low blood pressure of high plasma osmolarity stimulates the hypothalamus to release ADH to the posterior pituitary gland.

        c. From the posterior pituitary, ADH is released into the bloodstream, where it reaches the target tissue in the nephron.

   3. The **renin–angiotensin system** is another important hormonal system.

        a. The **juxtaglomerular apparatus** is a group of cells found where the afferent arteriole approximates the distal tubule (Figure 13-4).

        b. Stimulation by the **sympathetic nervous system** in times of low blood pressure, dehydration, or stress causes cells of the juxtaglomerular apparatus to release the enzyme **renin.**

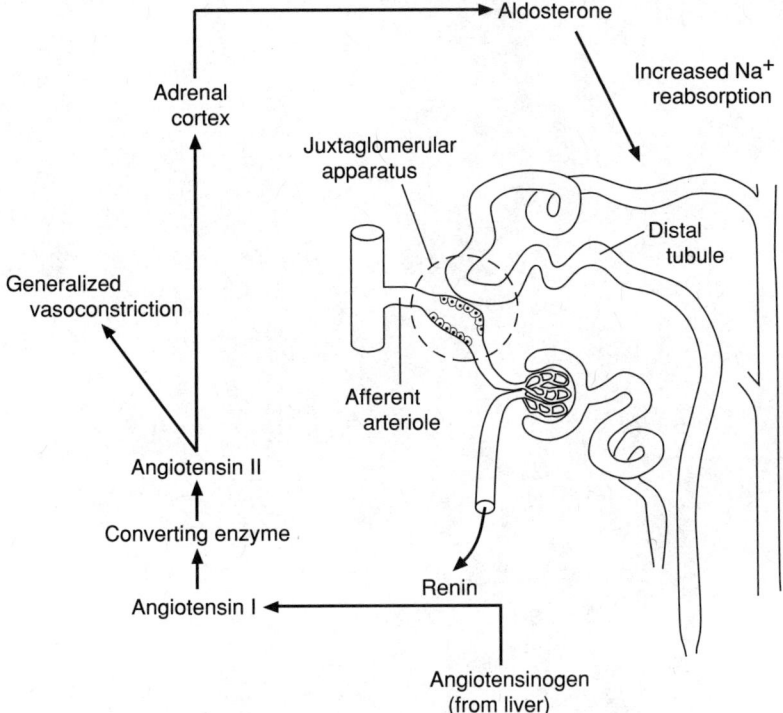

Figure 13-4. The renin–angiotensin system

    c.   Renin enters the circulation, where it converts a serum protein that is made in the liver, **angiotensinogen,** to a decapeptide called **angiotensin I.**

    d.   Angiotensin I passes through the circulation. When it enters the lung, it encounters a converting enzyme, which removes two amino acids and leaves an eight-residue peptide, or **angiotensin II.** Angiotensin II has two important roles.

        (1)   Angiotensin II is a **strong vasoconstrictor,** thus it increases blood pressure.

        (2)   It also **stimulates secretion of aldosterone,** which retains sodium and thereby increases blood pressure.

The Muscles and the Skeletal System

I. Muscle System

A. Functions

1. **Controlled movement and posture.** Muscle cells, like neurons, can propagate action potentials. However, unlike neurons, muscle cells can also contract and generate force. This characteristic enables the body to control the movement of body parts and posture.

2. **Muscle pairs.** Muscles tend to contract in groups rather than in isolation. Because the force of a muscle contraction can generate a pull but not a push, muscles tend to be arranged in antagonistic pairs.

   a. **Flexors** bend parts of the body across joints, and **extensors** straighten them.

   b. **Abductors** move limbs away from the body's central axis, and **adductors** move them toward the central axis.

   c. **Levators** raise body parts, and **depressors** lower them.

   d. **Pronators** move parts of the body downward and backward, and **supinators** move them upward and forward.

   e. **Sphincters** close hollow body parts, and **dilators** enlarge them.

B. Structural Organization (Figure 14-1)

1. A **muscle fiber** is a single muscle cell. Muscle fibers are multinucleated, and they contain contractile filaments in their cytoplasm. Muscle fibers are arranged parallel to one another (see Figure 14-1). For a detailed description of muscle cell structure and the mechanism of muscle contraction, refer to Chapter 7.

2. **Tendons** attach muscles to bones. The point of attachment that remains relatively fixed during contraction is called the **origin,** and the more mobile attachment site is termed the **insertion** (Figure 14-2).

C. Types of Muscle

There are three types of muscle: **skeletal, cardiac,** and **smooth** (Table 14-1).

D. Nervous Control

1. **Muscles** are innervated by both **sensory and motor nerve fibers.** Motor innervation can be somatic or autonomic, depending on the type of muscle (see Chapter 8).

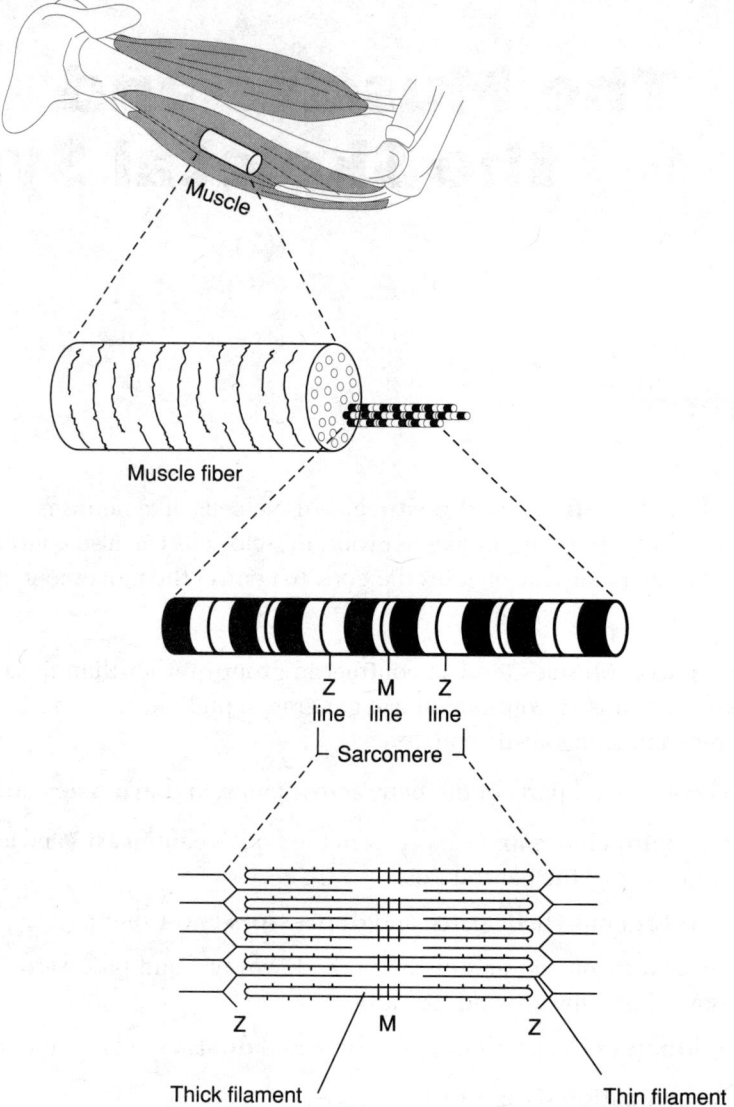

Figure 14-1. The structure of skeletal muscle.

2. **Transverse tubule** carries action potentials deep into the muscle tissue. Therefore, when an action potential from the nervous system causes depolarization of muscle fibers, the reaction of the muscle (i.e., contraction) can be better coordinated and synchronous because the depolarization is centrally located.

II. Skeletal System

A. Function

The human skeletal system is an **endoskeleton,** that is, it is internal, in contrast to the **exoskeleton** of insects, which is external. The function of the skeletal system is to:

1. **Support the structures** of the body

2. **Provide attachment sites** for muscle fibers

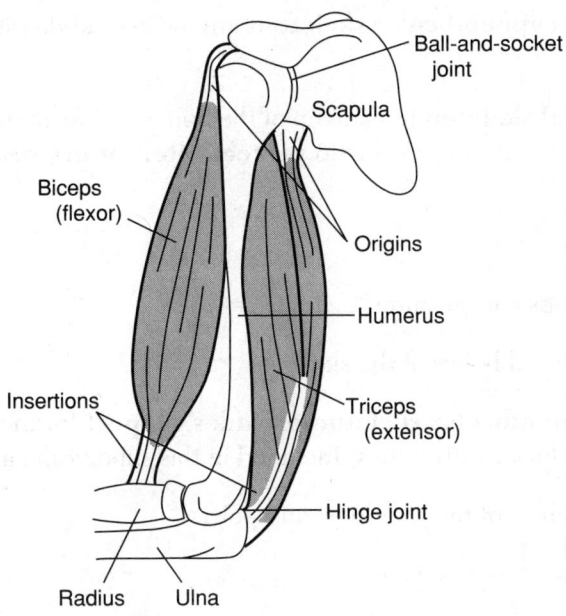

Figure 14-2. Anatomy of the upper arm.

3. **Form blood cells**
4. **Store inorganic ions** (i.e., calcium, phosphorus)
5. **Protect internal organs** from injury

B. **Structure**
   1. **Basic concepts** of the skeletal system are as follows:
      a. The human skeletal system is made up of both **cartilage** (see II D) and **bone**.
      b. A **joint** is the site where two or more bones join, or **articulate**.
      c. **Ligaments** attach bones to one another.

TABLE 14-1. Comparison of Muscle Type-Specific Characteristics

Characteristics	Skeletal	Cardiac	Smooth
Location	Attached to bones or cartilage via tendons	Heart tissue	Lining of hollow organs (e.g., gut, blood vessels)
Cell shape	Large, elongated, cylindrical, and blunt ends	Short, branching, and arranged end to end	Spindle shaped, flattened, and pointed ends
Number and location of nuclei	Multinucleated/peripheral	Multinucleated/central	One/central
Striations	Present	Present	Absent
Blood supply	Good	Rich	Fair
Control of contraction	Voluntary	Involuntary	Involuntary
Speed of contraction	Most rapid	Intermediate	Slowest
Ability to remain contracted	Least	Intermediate	Greatest

2. The **axial and appendicular skeletons** are the two subdivisions of the human skeletal system.

   a. The **axial skeleton** is made up of the bones that form the trunk and skull of the body. Its main function is to **protect internal organs.** The axial skeleton includes:

      (1) Vertebral column

      (2) Ribs and sternum

      (3) Fused bones of the skull

   b. The **appendicular skeleton provides support** for the appendages of the upper and lower extremities. Included in this subdivision are:

      (1) Bones of the shoulders and pelvis

      (2) Arms, wrists, hands

      (3) Legs, ankles, feet

3. **Several types of bones are based on shape.**

   a. **Long bones** include the femur, tibia, radius, and ulna.

   b. **Short bones** include the wrist and ankle bones.

   c. **Flat bones** include the cranial bones, sternum, and ribs.

   d. **Irregular bones** include the vertebrae.

4. There are several major **types of joints.**

   a. **Ball and socket joints** include the hip and shoulder joints, which give **many types of movement.**

   b. **Hinge joints** include the elbow and knee, which allow **motion in one plane only.**

   c. **Condylar joints** include the jaw joint, which allows **motion in two planes.**

   d. **Rotating joints** include the radius and ulna, which **only allow rotation.**

C. **Bone**

   1. **Structure.** Bone is a specialized type of connective tissue (see Chapter 7). Bone tissue is composed of cells that are embedded in a matrix of organic fibers and inorganic ions.

      a. **Cells**

         (1) **Osteoblasts** are cells in the active synthesis of bone matrix. They are stimulated by growth hormone.

(2) **Osteocytes** are dormant osteoblasts that have surrounded themselves with matrix. They can be reactivated when a bone is injured.

(3) **Osteoclasts** are multinucleated cells that remodel bone and release inorganic (i.e., calcium, phosphorus) and organic components. They are stimulated by parathyroid hormone.

b. **Matrix**

(1) **Organic** matrix includes collagen fibers (mostly type I) and glycoproteins.

(2) **Inorganic** matrix consists of ions, of which the most abundant form is calcium phosphate in a crystalline form called **hydroxyapatite.**

2. **Types**

a. **Primary (immature) bone** is the first bone formed in the fetus. It is also found during bone repair (i.e., it is temporary). Primary bone has irregular arrays of collagen and a low inorganic content.

b. **Mature bone**

(1) **Spongy bone** is found in the marrow space of long bones (Figure 14-3). Characteristics of the structure of spongy bone include the following:

(a) Interconnected trabeculae forming meshwork

(b) Bone marrow present in the spaces of the meshwork

(2) **Compact bone** is the hard, outer layer of bone (see Figure 14-3C). The structure of compact bone, which is solid and mature, is made up of numerous concentric units called **osteons** or **haversian systems.** Each osteon has central blood vessels and concentric rings of osteocytes embedded in a bone matrix.

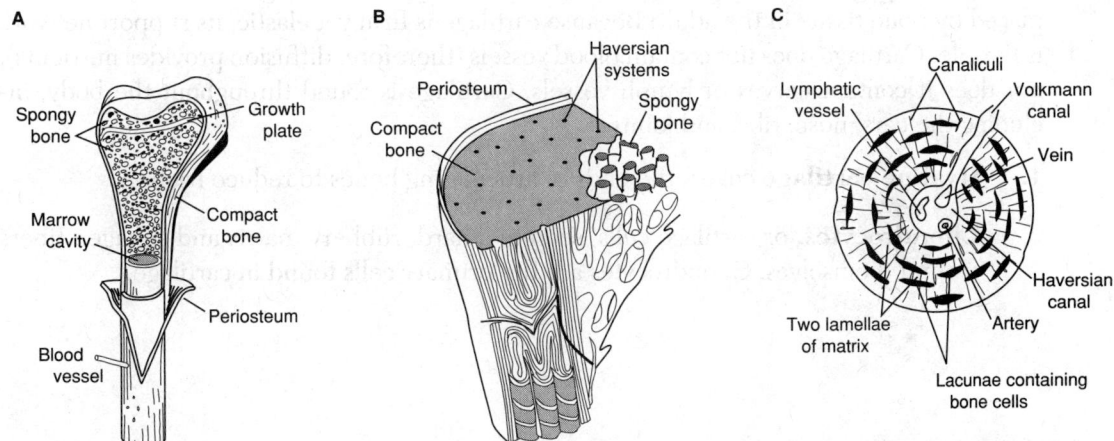

Figure 14-3. Bone anatomy. In **A)**, note the periosteum, a fibrous outer layer over the bone. The compact bone is the outer layer of bone. The spongy bone is the inner layer. In **B)**, the osteons (haversian systems), which are the small, functional units of mature bone, are shown. In **C)**, the microstructure of an osteon is demonstrated. Note the lacunae containing osteocytes.

- (a) **Volkmann's canals** carry blood vessels into the bone tissue from the outside.
- (b) **Haversian canals** enclose blood vessels that bring nutrients to the osteon or haversian system.
- (c) **Lacunae** are spaces in the bone matrix that contain the cell bodies of osteocytes.
- (d) **Canaliculi** are networks of passageways between the lacunae that allow osteocytes to communicate and exchange nutrients.

3. **Bone formation**

   a. **Intramembranous bone** forms directly from connective tissue without a cartilage precursor. Examples include the bones of the skull.

   b. **Endochondral bone** forms from a cartilage precursor. The cartilage is reabsorbed and replaced by bone tissue. Examples include the long bones of the body (e.g., humerus, femur).

4. **Bone growth**

   a. Growth of the **periosteum** increases the **width** of bone. The periosteum is a fibrous layer of connective tissue that covers bone. It contains cells that can mature into osteoblasts or osteoclasts.

   b. Growth of the **epiphyseal plates** increases the **length** of bone. Epiphyseal plates (growth plates) are cartilaginous structures between the **diaphysis** (shaft) and **epiphysis** (end) of a bone. These structures function through puberty, then calcify and no longer function.

D. **Cartilage**

During embryogenesis, cartilage is the structure of support, although much of it is later replaced by bone tissue in the adult. Because cartilage is firm yet elastic, its support network is flexible. Cartilage does not contain blood vessels (therefore, diffusion provides nutrients), nor does it contain nerves or lymph vessels. Cartilage is found throughout the body, including the ears, nose, ribs, and joints.

1. **Hyaline cartilage** covers the ends of articulating bones to reduce friction.

2. **Chondrocytes,** or cartilage cells, secrete a hard, rubbery matrix and collagen fibers around themselves. Chondrocytes are the primary cells found in cartilage.

The Respiratory System

15

I. Function and Basic Anatomy of the Airways and Lungs

A. Function

Animals use oxygen and produce carbon dioxide during the process of cellular respiration. Organisms that are smaller than 0.5 mm in diameter can use diffusion alone for gas exchange. As organisms become larger, diffusion distances are increased and the ratio of surface area to volume is reduced. A circulatory system and respiratory system has evolved in larger animals to facilitate gas exchange.

B. Anatomy (Figure 15-1)

1. **Structures of the airways** include the nose, mouth, pharynx, larynx, and trachea. Air enters the **nose** or **mouth** and moves into the pharynx. The **pharynx** is the upper throat and is located behind the nasal passages and behind the tongue. Air then moves into the **larynx,** the area of the lower throat where the vocal cords are located. Air must pass between the two vocal cords to enter the **trachea.**

   a. The trachea is a rigid tube made of cartilaginous rings.

   b. The trachea is rigid to prevent collapse from the negative pressures generated during inspiration.

2. **Regions of the airways**

   a. The **main bronchi** are the two main branches of the trachea. There is a left and right main bronchus.

   b. **Bronchioles** are the smaller sets of branches created each time the main bronchus divides (which is more than 20 times). Most of the early branches have a cartilage-supported wall and are stiff. The later branches have smooth muscle in the wall, which allows for dilation or contraction.

   c. The **alveolus** is the smallest structure at the end of the branched network. Alveoli are lined with single cells and are spherical chambers where most gas exchange occurs (Figure 15-2). Gas exchange largely occurs across the cell membranes of the alveoli. There are more than 300 million alveoli in adult human lungs.

   d. **Alveolar sacs** rise from clusters of alveoli, and alveolar sacs communicate with one another via **alveolar ducts.**

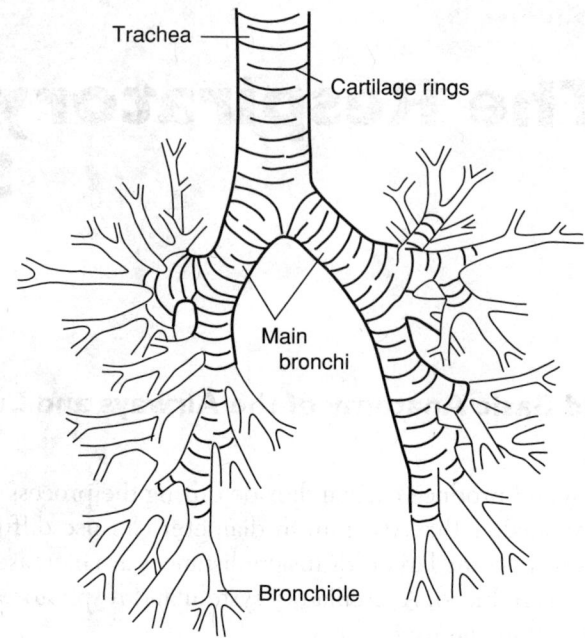

Figure 15-1. The airways of the lung.

        e.   The **terminal bronchiole** carries the air to the alveolar network.

        f.   An **acinus** is an alveolar network supplied by one terminal bronchiole.

    3.   **Surfactant** is a lipid material that lines alveoli. This substance decreases the surface tension of the spherical alveoli and reduces the likelihood of alveoli collapsing. Some premature infants have respiratory distress and alveolar collapse because they do not produce adequate amounts of surfactant.

II. Hemoglobin, Gas Exchange, and Respiratory Equations

A. Hemoglobin

Hemoglobin is a respiratory pigment that is the main oxygen carrier in the blood.

    1.   **Structure.** Hemoglobin is a large molecule with a molecular weight of 68,000.

        a.   Each molecule contains **four subunits,** and each subunit contains **a polypeptide chain** and **heme,** which is an iron-containing prosthetic group.

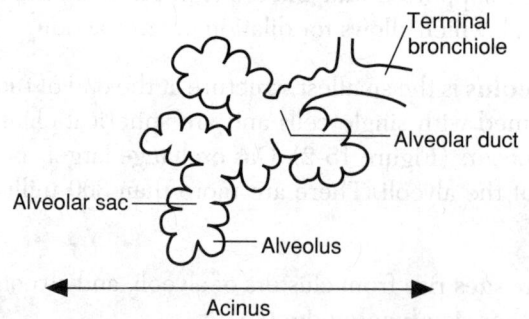

Figure 15-2. The smallest components of the respiratory tree.

b. **Polypeptide chains** are either **alpha** or **beta.** Each hemoglobin molecule contains two alpha and two beta chains. Each chain has a heme moiety associated with it.

2. **Binding properties**. Each subunit can bind one molecule of oxygen. Thus, **one hemoglobin molecule can bind four oxygen molecules.** The binding of the first oxygen molecule makes the binding of the next oxygen molecule easier, which is referred to as **positive cooperativity.**

3. **Myoglobin versus hemoglobin. Myoglobin** is a storage pigment for oxygen located in muscle tissue. The major structural difference between myoglobin and hemoglobin is that myoglobin has only one subunit containing one heme group. Myoglobin can bind only one oxygen molecule, so it does not demonstrate positive cooperativity.

4. **Hemoglobin–myoglobin dissociation curves** (Figure 15-3). The dissociation curves show partial pressure of oxygen (PO_2) on the x-axis and the percentage of hemoglobin or myoglobin saturated with oxygen on the y-axis. Room air contains approximately 150 mm Hg of oxygen. Note that the curve for myoglobin is hyperbolic, whereas the curve for hemoglobin is **sigmoid.**

   a. **Myoglobin** has one heme group and binds only one oxygen molecule, so it quickly saturates with oxygen.

   b. **Hemoglobin** is slower to take up its first oxygen molecule, and the binding of each subsequent oxygen molecule makes further binding easier (i.e., positive cooperativity), causing the upturn and sigmoid shape of the curve.

   c. **Both curves flatten** as the pigments become fully oxygen bound.

5. **Factors affecting the dissociation curves.** The **oxygen affinity of hemoglobin** is labile. Affinity can be increased by decreasing temperature, basic blood pH, or low levels of diphosphoglycerate (DPG). This causes the hemoglobin–oxygen dissociation curve to shift to the left. Oxygen affinity of hemoglobin can be reduced by the following factors, which shift the curve to the right (Figure 15-4):

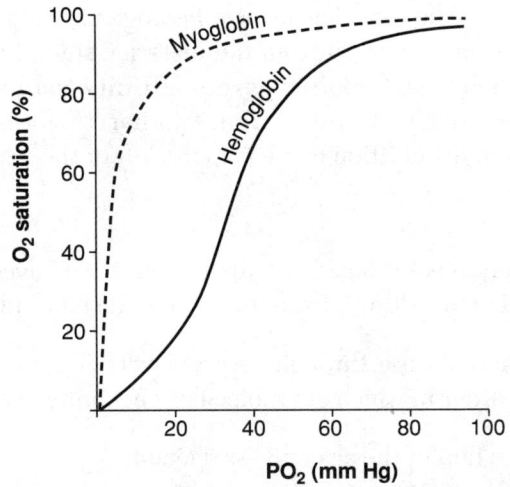

Figure 15-3. The dissociation curves for hemoglobin and myoglobin.

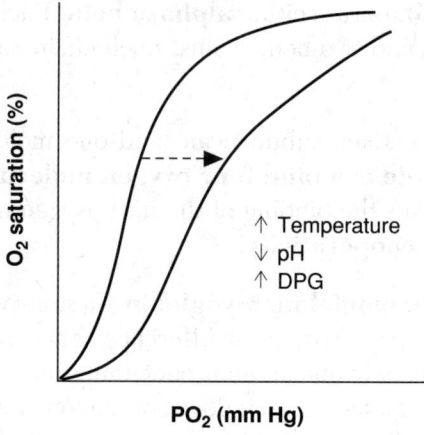

Figure 15-4. The effect of increased temperature, decreased pH, and increased diphosphoglycerate (DPG) levels on the hemoglobin–oxygen dissociation curve. Note the shift to the right.

    a. **Increases in temperature** make it more difficult to load hemoglobin with oxygen. Hemoglobin also gives up oxygen more easily at high temperatures. At high body temperatures, the rate of metabolism is higher, and the need for oxygen at the tissue level is greater.

    b. When **carbon dioxide concentration is increased,** carbon dioxide is converted to carbonic acid and **blood pH decreases.** A decrease in blood pH makes it more difficult to load hemoglobin with oxygen and makes oxygen unloading to the tissues easier. The effect of pH on hemoglobin–oxygen affinity is called the **Bohr effect.** The enhanced unloading of oxygen by hemoglobin is useful in periods of exercise, stress, or increased blood acidity.

    c. **Increases in DPG.** Increases in the products of anaerobic metabolism decrease the affinity of hemoglobin for oxygen and increase the unloading of oxygen to the tissues. DPG is produced in anaerobic metabolism (glycolysis) and has a direct effect on hemoglobin. When anaerobic conditions exist, DPG levels rise, and oxygen is given up more easily by hemoglobin. This makes more oxygen available at the tissue level to provide for oxidative metabolism.

6. **Fetal hemoglobin.** The fetus has its own hemoglobin, and there must be a driving force for the diffusion of oxygen from the maternal side of the placenta to the fetal side. Fetal hemoglobin has a **higher oxygen affinity than adult hemoglobin;** this is what enhances oxygen transfer from mother to fetus. Therefore, the fetal hemoglobin–oxygen dissociation curve is to the left of the adult curve.

B. Gas Exchange

1. **Diffusion.** Oxygen must diffuse from the lumen of the alveolus to a red blood cell, and carbon dioxide must diffuse from the red blood cell or plasma to the alveolus.

2. **Layers.** Gases must diffuse through several layers (Figure 15-5). These layers are listed here in the order in which gas molecules encounter the structures:

    a. Fluid surface film of the alveolus (surfactant)

    b. Alveolar lining (membrane)

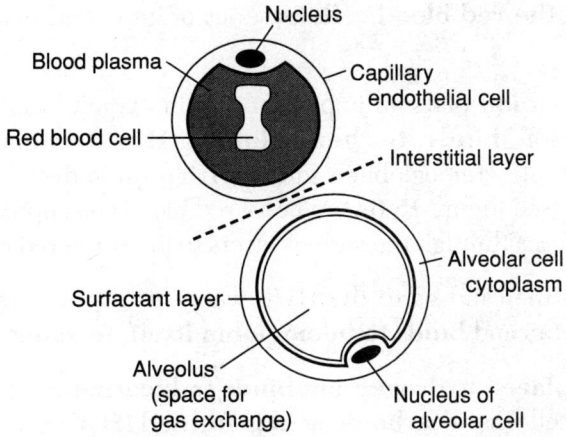

Figure 15-5. Schematic diagram showing the microscopic anatomy of gas exchange (in cross section).

    c. Interstitial layer

    d. Capillary endothelial lining

    e. Blood plasma

    f. Wall of the red blood cell (membrane)

C. Equations for Oxygen Loading and Unloading (Figure 15-6)

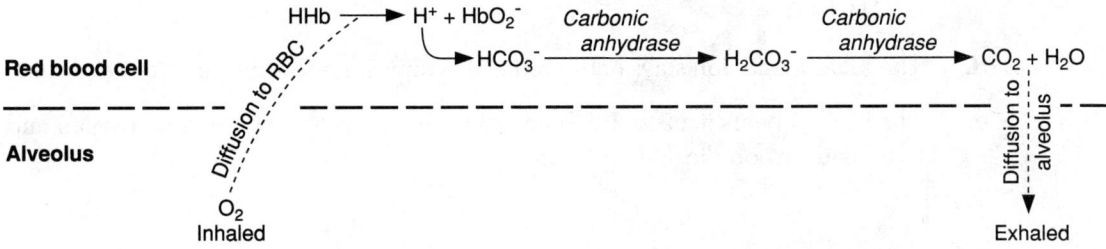

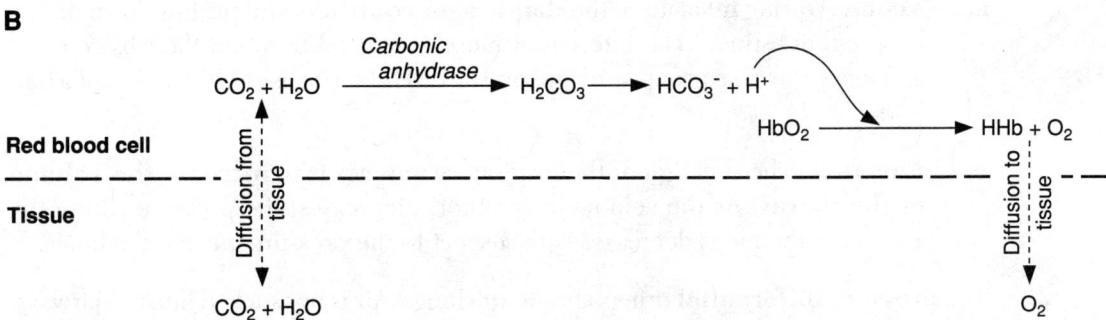

Figure 15-6. Equations showing oxygen loading and unloading at **A)** the red blood cell–alveolus interface and **B)** the red blood cell–tissue interface.

1. **Reactions at the red blood cells.** A series of important reactions occur at the red blood cell.

    a. **Hydrogen ion** plays an important role in oxygen loading and unloading. Hydrogen ion **binds to hemoglobin (Hb) to form HHb** (un-ionized hemoglobin). Hemoglobin is in the HHb form in the red blood cells that enter the lung (see Figure 15-6A). When a red blood cell approaches an alveolus, **oxygen** diffuses through the various layers to enter the red blood cell.

    b. **Oxygen** then moves to the HHb where it displaces the hydrogen ion from hemoglobin and **binds to hemoglobin itself, forming HbO_2**

    c. The **displaced hydrogen ion binds to bicarbonate ion** in the cytoplasm of the red cell where **carbonic acid is formed (H_2CO_3).**

    d. An enzyme called **carbonic anhydrase** then **converts carbonic acid to CO_2 and H_2O.** The carbon dioxide and water diffuse out of the red blood cell where they are expelled (i.e., exhaled).

2. **Reverse reactions at the tissues** (see Figure 15-6B)

    a. HbO_2 is the form of hemoglobin that carries oxygen to the tissues.

    b. At the tissues, CO_2 and H_2O, products of oxidative metabolism, diffuse into the red cell. Carbonic anhydrase converts them to carbonic acid. Carbonic acid dissociates into hydrogen ion and bicarbonate ion.

    c. The hydrogen ion displaces the oxygen from the HbO_2 and binds to Hb, forming HHb.

    d. The bicarbonate ion stays in the red cell cytoplasm or diffuses out to the plasma.

    e. The cycle repeats itself as HHb moves to the lungs to pick up more oxygen and to unload carbon dioxide and water.

III. Mechanics of Breathing

A. **Inhalation and Exhalation**

1. **Inhalation (inspiration)** requires energy and is an active process.

    a. **Action.** During inhalation, the **diaphragm contracts** and pushes down on the liver and intestines. The intercostal muscles located between the ribs contract and move the rib cage upward and outward. The action is similar to that of a bellows.

    b. **Function.** The function of the inspiratory muscles is to **increase the volume of the thorax.** As the volume of the thorax increases, the pressure within the airways of the lungs decreases with respect to the pressure outside the lungs.

2. The **pressure differential** brings air into the lungs. Air is not sucked into the airways; air is attracted into the airways by the pressure gradient created by increased thoracic volume.

3. **Exhalation (expiration)** is usually a passive process. (However, exhalation can be forceful when neck muscles are contracted and a different set of intercostal muscles are used.)

   a. **Action.** The diaphragm relaxes and moves superiorly. The liver and intestines push back up on the base of the lungs. The intercostal muscles relax, and the rib cage moves down and in.

   b. **Function.** The combination of muscle relaxation and recoiling of the lungs decreases the volume of the thorax, creating a higher pressure inside the airways than the pressure of air outside the body. Air moves down its pressure gradient and is exhaled.

B. **Lung Spaces and Pressures**

   1. The **pleural space** of the thorax is shown in Figure 15-7. This space is filled with a small volume of pleural fluid, and there is a separate pleural lining for each lung. The pleural space is the potential space between the following two lung linings:

      a. **Parietal pleura**—the lining on the inside of the rib cage

      b. **Visceral pleura**—the lining directly on top of the lung

   2. The lungs have a **continual elastic tendency** to collapse and pull away from the chest wall. This is caused, in part, by the surface tension of the alveoli. The elastic content of the lung also contributes to its recoil.

   3. The **total recoil tendency** of the lungs can be measured by the amount of negative pressure in the intrapleural space required to prevent collapse of the lungs. This pressure is the **intrapleural pressure,** which is approximately −4 mm Hg. When the airways are open to atmospheric pressure, a negative pressure of approximately −4 mm Hg is required to keep the lungs expanded to normal size and to overcome the surface tension and elastic fiber forces.

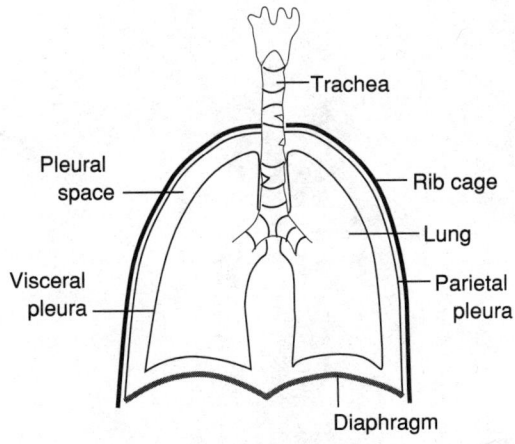

Figure 15-7. Schematic diagram of the lung linings and the pleural space.

IV. Thermoregulation

A. Increases in Lung Ventilation Increase Heat Loss from the Respiratory System

1. **When air is inhaled,** it is warmed and humidified.

2. **When air is exhaled,** it is cooled in the respiratory passages, conserving some potentially lost heat. However, when the rate of ventilation increases, less time is available to reabsorb heat energy.

B. Mammals and Birds Control Heat Loss via the Respiratory System to Regulate Body Temperature

To increase heat loss, mammals breathe through the mouth and hyperventilate. For example, a "panting" dog inhales through the nose and exhales through the mouth. The exposed tongue encourages water evaporation and, consequently, heat loss.

V. Protective Mechanisms Against Disease and Particulates

A. The **upper respiratory tracts** are **lined with a ciliated mucosa,** which contains mucous cells.

1. **Mucus** is produced in the upper airways and is moved by cilia toward the mouth. The mucus is swallowed when it arrives in the throat.

2. The function of mucus is to **trap small particulate matter** such as dust, dirt, bacteria, and pollen.

B. The **nasal cavities** also trap particulates and are the first line of defense against particle inhalation.

1. **Nasal hairs and nasal mucus** trap many larger particles.

2. The small particles that travel down the bronchial tree may undergo **phagocytosis** by macrophages.

C. The **lung** is protected by the immune system; it has **resident macrophages** and a constant flow of **circulating immune cells** to protect it.

Skin 16

I. Structure of the Skin

A. The **epidermis,** which is derived from ectoderm, is made up of a **stratified squamous epithelium** that expresses intermediary filaments of keratin (see Chapter 7). This epithelium includes several layers of cells, with the bottom layer attached to a basement membrane that divides the epidermis from the underlying dermis (see I B).

1. As **dead cells** are removed from the skin's surface, they are **replaced by cells underneath.** The rate of cell replacement equals the rate of sloughing so that a constant epidermal thickness is maintained, although the overall thickness varies in different parts of the body.

2. The **cells of the basal layer** just above the basement membrane **divide rapidly and continuously.** As the cells migrate up toward the body surface, they differentiate and begin to express the protein keratin. **Keratin** contributes to the mechanical strength and flexibility of the skin.

3. Because the epidermis does not contain blood vessels, **the differentiating cells become nutrient deprived** and, therefore, less metabolically active. As the cells approach the surface of the skin, they lose their nuclei, die, and are shed by physical abrasion.

B. The **dermis** is the **connective tissue layer** between the epidermis and the hypodermis (see I C). Like other connective tissues (see Chapter 7), the dermis is composed of collagen, reticulin, and elastic fibers. The fiber lattice acts as a scaffolding for the connection of the basement membrane as well as for skin-associated structures (e.g., hair, glands). Unlike the epidermis, the dermis has extensive capillary networks that nourish the dermis directly and the epidermis indirectly through the diffusion of nutrients and gases across the basement membrane. Many **skin-associated structures** are embedded in the dermis (Figure 16-1).

1. Glands

    a. **Sebaceous glands** produce **sebum,** which is an oily substance that coats the exterior surface of the skin, slows the evaporation of water, and has antimicrobial effects. Sebum is secreted from ducts into the shaft of the hair follicle, thereby gaining access up through the epidermis to the surface of the skin.

    b. **Sweat glands** secrete metabolic waste products, salt and other ions, and water.

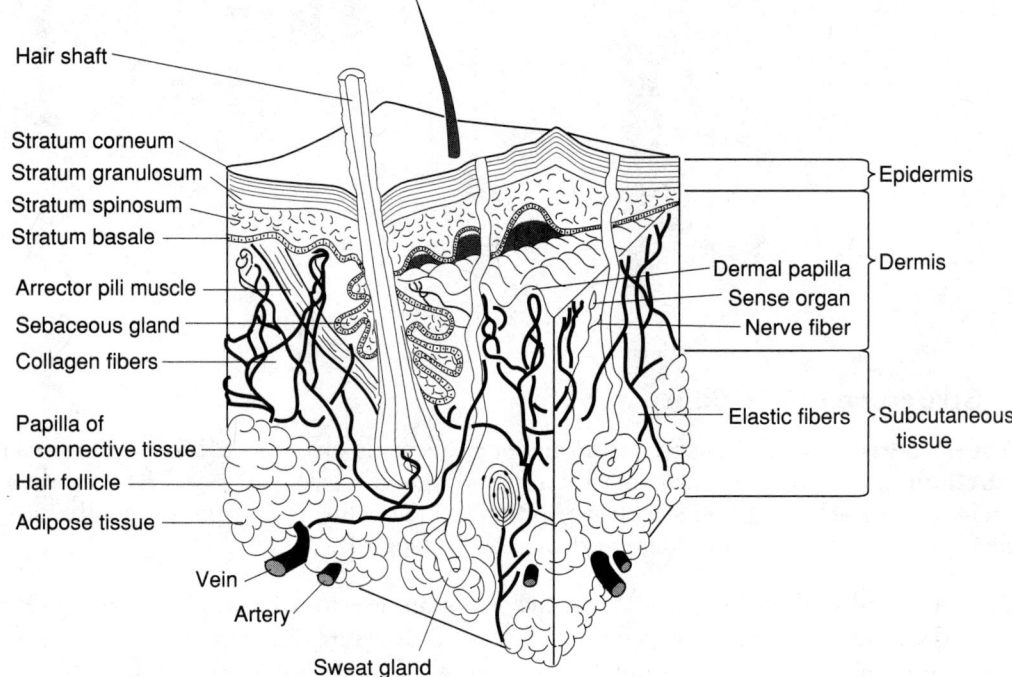

Figure 16-1. Microscopic structure of the skin. (Reproduced with permission from Ville CA, Solomon EP, Davis PW: Biology. Philadelphia, Saunders College Publishing, 1985, p 662.)

2. **Hair follicles** begin in the dermal layer and extend up through the epidermis from which they are originally derived. Hairs are closely associated with sebaceous glands as well as several sensory structures (see Chapter 8). To make the hair stand up, the erector pili muscles contract, which pulls the hair shaft perpendicular to the surface of the skin.

C. **Hypodermis** is a layer of **subcutaneous tissue** below the dermis. In the hypodermis, **adipocytes,** or fat cells, are numerous. The adipocytes create a layer of subcutaneous fat that helps to insulate the body. Capillary networks tend to be sparse because adipocytes are less metabolically active than most cell types.

II. Function of the Skin

A. **Homeostasis and Osmoregulation**

1. **Homeostasis** is the tendency of organisms to **maintain an internal steady state,** despite changes in the external environment. In all organisms, a dynamic yet balanced internal state must be maintained. To accomplish this goal, metabolic processes must be monitored and constantly adjusted. These self-regulating control measures are exquisitely sensitive to small perturbations and are remarkably efficient in preventing excessive loss of time and energy.

2. **Osmoregulation** is the process that **maintains the water content of the body** in addition to **regulating** the concentration and distribution of the various ions.

a. **Sweat glands.** Although the kidneys are the primary center for osmoregulation, the sweat glands in the skin excrete between 5% and 10% of the body's metabolic wastes, including salt and urea. Perspiration can vary from 0.5 L on a temperate day to 2–3 L on a hot day. The loss of both water volume and salt aids in the control of body osmolarity.

b. **Protection against dehydration.** The skin plays another pivotal role in osmoregulation by protecting the body against dehydration. Human skin is impermeable to water and is, therefore, a barrier to water exchange with the environment. In other animals, such as amphibians, the skin is permeable to water, which allows those animals to osmoregulate directly with their environment.

B. **Thermoregulation**

The regulation of body temperature is an example of a homeostatic operation. Although some heat is dispelled through respiration, defecation, and urination, **90% of total heat loss is through the skin.**

1. **When body temperature increases,** cells in the hypothalamus sense the increase and send out nerve impulses to compensate.

    a. **Increased sweat output.** Sympathetic neurons (see Chapter 8) stimulate sweat glands to increase their output of sweat. The subsequent evaporation of the sweat from the surface of the skin helps to lower body temperature. Heat from the skin surface is dissipated by the conversion of sweat into water vapor.

    b. **Vasodilation.** Other nerve impulses travel to capillaries in the skin and cause them to vasodilate. This process increases the volume of blood being delivered to the surface of the skin so that more heat can radiate away.

2. **When body temperature decreases,** the hypothalamus senses the heat loss.

    a. **Vasoconstriction.** The hypothalamus sends nerve impulses to induce vasoconstriction of skin capillaries, which reduces heat loss through radiation.

    b. **Muscle contraction (shivering).** In more extreme cases of heat loss, nerve impulses stimulate skeletal muscles to rapidly contract. The shivering generates heat without causing a great loss of energy because the amount of body movement is minimal.

C. **Physical Protection**

The skin acts as a physical barrier between the body and the external environment. This physical separation serves several purposes.

1. Entry of infectious agents is prevented.

2. Damage from chemical toxins, extreme temperature fluctuation, and ultraviolet rays is reduced.

3. The body is shielded from abrasive forces encountered during movement.

The Reproductive System

I. Male Reproductive System

A. **Genitalia and Gonads**

The male reproductive system is composed of the structures shown in Figure 17-1. The overall function of this system is to produce mature, active sperm and to deliver them outside the body.

1. **Passage of sperm. Spermatogonium** are the male gametocytes (Figure 17-2).

   a. These progenitor cells arise in the **seminiferous tubules of the testes.**

   b. Spermatogonium are nourished by **Sertoli cells.** Development of spermatogonium takes several days, with the mature spermatozoa being concentrated and stored in the lower **epididymis** until ejaculation.

   c. Upon arousal, the penis becomes engorged with blood, and spermatozoa are propelled up into the **vas deferens** during ejaculation.

   d. From the vas deferens, the spermatozoa pass sequentially through the **urethra** and exit the body.

2. **Semen,** or ejaculate, contains **spermatozoa** and **seminal fluid.** One ejaculation produces 3–4 mL of semen and contains 300–500 million spermatozoa.

   a. The seminal fluid is **secreted by the male accessory sex glands,** of which there are three types.

      (1) The **prostate gland** is the largest seminal fluid–producing gland. There is one prostate gland.

      (2) The **seminal vesicles** are paired glands that secrete a component of semen.

      (3) The **bulbourethral (Cowper) glands** are paired, and they produce a mucoid secretion that is a component of semen.

   b. The fluid and its contents serve several purposes.

      (1) The **fluid** provides a vehicle for the **transport of the spermatozoa** out of the body.

      (2) **Fructose** in the fluid provides the spermatozoa with an energy source.

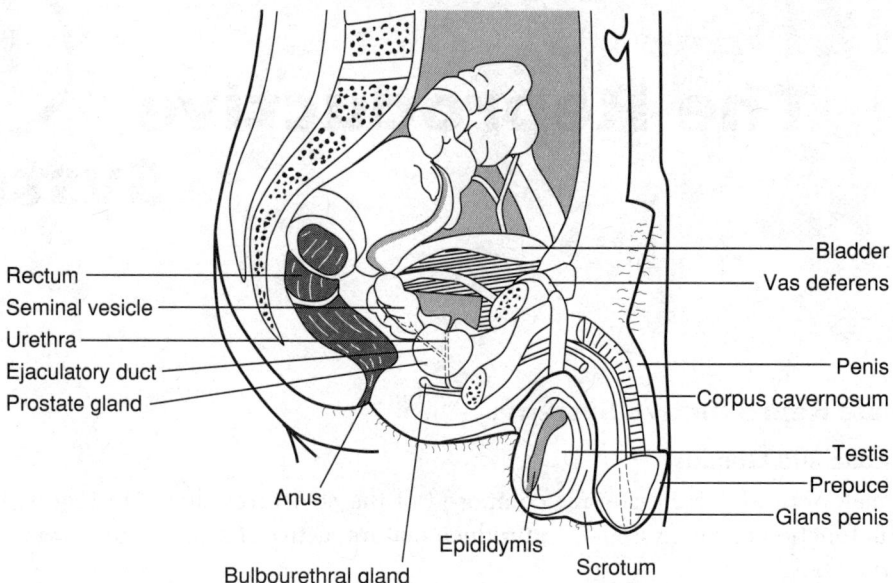

Figure 17-1. Male reproductive anatomy. (Reproduced with permission from Ville CA, Solomon EP, Davis PW: *Biology*. Philadelphia, Saunders College Publishing, 1985, p 934.)

        (3) There may be **chemicals** in the fluid that protect the spermatozoa from the harsh environment of the uterus.

B. Spermatogenesis

    1. **Gametogenesis.** Gametogenesis is the production of sex cells, or **gametes** (i.e., spermatozoa in men; ova in women).

        a. **Gametes** contain only one set of chromosomes (i.e., 23 chromosomes) and are **haploid. (Somatic cells** are **diploid;** they contain 23 pairs of chromosomes, or 46 chromosomes, with one member of each pair being contributed by each parent.)

        b. The **reduction in the number of chromosomes** occurs during **meiosis** (see Chapter 19) as part of gametogenesis.

        c. When the egg and sperm unite **at fertilization, the diploid state is regained.**

    2. **Spermatogenesis.** In all vertebrates, spermatogenesis takes place in the **testes.**

        a. The **spermatogonia,** which are primitive, unspecialized germ cells, line the walls of the seminiferous tubules. Throughout embryonic development and before puberty, the spermatogonia divide mitotically to give rise to more spermatogonia. Only **after puberty** do the spermatogonia begin the process of spermatogenesis, which produces mature spermatozoa. Once spermatogenesis starts, it continues throughout a man's life.

        b. The **steps of spermatogenesis** are shown in Figure 17-3.

            (1) The process begins with the growth of the spermatogonia into larger cells called **primary spermatocytes.**

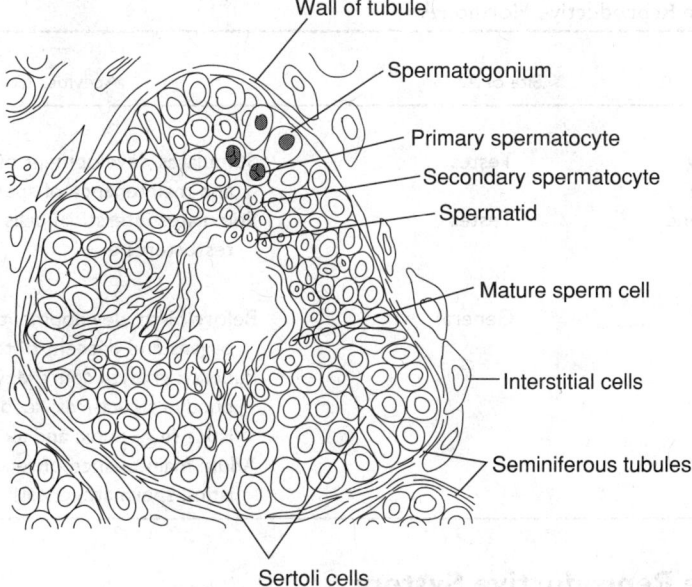

Figure 17-2. Seminiferous epithelium. Note that the immature cells are spermatogonium, and they are on the periphery of the tubule. As cells undergo spermatogenesis and mature, they move toward the center of the tubule and become mature sperm cells. (Reproduced with permission from Ville CA, Solomon EP, Davis PW: *Biology*. Philadelphia, Saunders College Publishing, 1985, p 935.)

      (2)   During meiosis I, the primary spermatocytes give rise to two **secondary spermatocytes,** each of which undergoes a second division, meiosis II, to yield four **spermatids.**

      (3)   All four spermatids mature into fertile **spermatozoa.**

  3. **Hormonal control.** Spermatogenesis is regulated by secretions from the **anterior pituitary gland.**

     a. **Luteinizing hormone (LH)** stimulates the **interstitial cells of Leydig** to produce **testosterone,** which then aids in the process of spermatogenesis.

     b. Maintenance of spermatogenesis is assisted by **follicle-stimulating hormone (FSH)** and **growth hormone (GH).** For a summary of the male reproductive hormones, see Table 17-1 and Chapter 9.

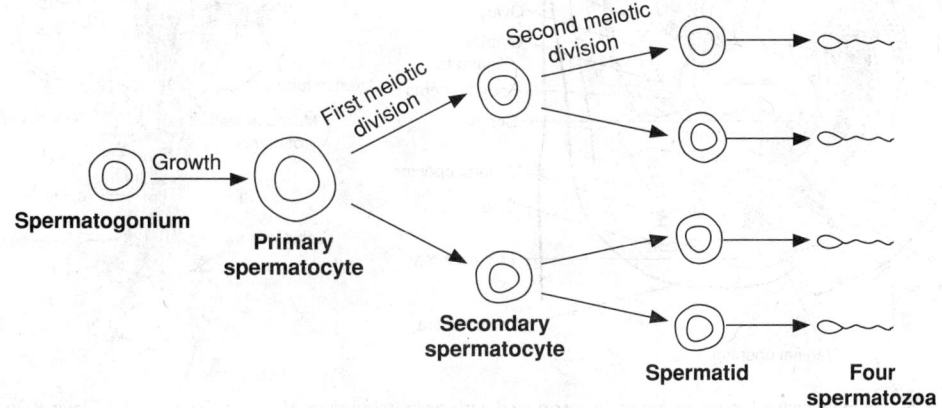

Figure 17-3. Spermatogenesis.

TABLE 17-1. Male Reproductive Hormones

Endocrine Gland and Hormones	Site of Action	Activities Controlled
Anterior Pituitary		
Follicle-stimulating hormone (FSH)	Testes	Stimulates development of seminiferous tubules and maintenance of spermatogenesis
Luteinizing hormone (LH)	Testes	Stimulates interstitial cells of Leydig to release testosterone
Testes		
Testosterone	General	Before birth: development of primary sex organs and descent of testes into scrotum Puberty: growth spurt; development of reproductive structures; secondary sex characteristics (e.g., pubic and axillary hair, deep voice) Adult: maintenance of secondary sex characteristics; spermatogenesis

II. Female Reproductive System

A. Genitalia and Gonads

The structures of the female reproductive system are shown in Figure 17-4.

1. **The function of the female reproductive system is twofold:**

   a. To produce fertile eggs or ova

   b. To provide an environment capable of supporting embryonic development if an egg becomes fertilized

2. The **process of fertilization** is as follows: during intercourse, the man's penis is inserted into the woman's **vagina.** Upon ejaculation, spermatozoa travel up through the **cervix** into the **uterus** and ultimately reach the **fallopian tubes** or **oviducts.** In most

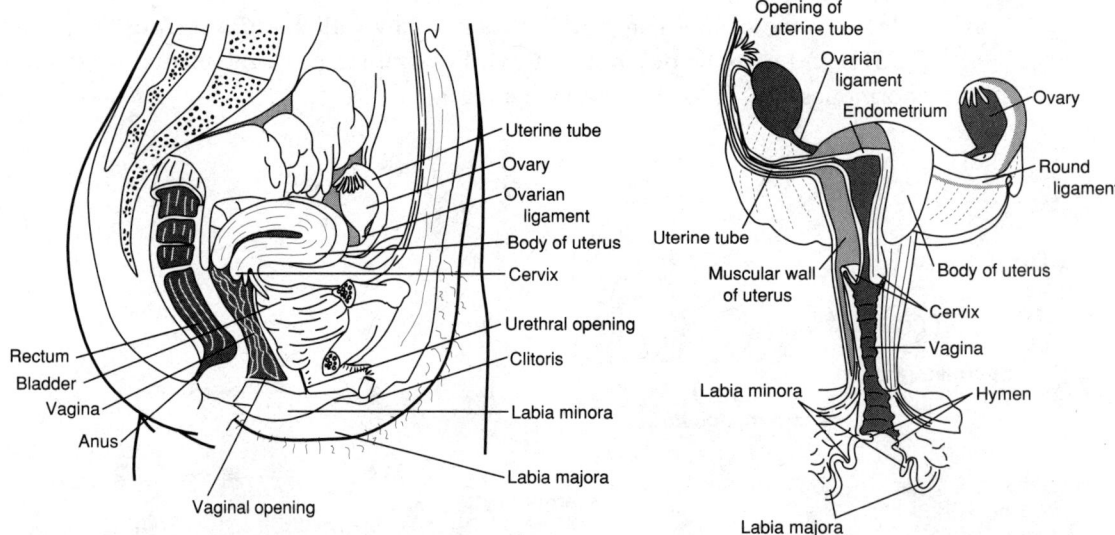

Figure 17-4. Female reproductive anatomy. (Reproduced with permission from Ville CA, Solomon EP, Davis PW: *Biology.* Philadelphia, Saunders College Publishing, 1985, p 936.)

TABLE 17-2. Spermatogenesis Versus Oogenesis

	Spermatogenesis	Oogenesis
Meiosis	Continuous	Arrested in meiosis I until ovulation; meiosis II does not occur unless there is fertilization
Time frame	Throughout lifetime	Maximum number at birth
Meiotic products	Four fertile spermatozoa	One mature ovum and up to three nonviable polar bodies
Division of cytoplasm	Equal	Unequal (large ovum and small polar bodies)
Site of development	Entirely within testes	Meiosis I within ovary; meiosis II wherever fertilized
Hormones	FSH and GH: maturation of spermatozoa LH: stimulation of Leydig cells to make testosterone	FSH: stimulates follicle cells to produce estrogen LH: stimulates corpus luteum to produce progesterone

pregnancies, fertilization takes place in the distal segment of the fallopian tube near the ovary.

B. Oogenesis

Table 17-2 compares oogenesis and spermatogenesis.

1. **Ova development**

    a. **Female embryos.** The **ova,** or eggs, begin development within the ovary from immature **oogonia** (Figure 17-5). Early in embryonic development, the oogonia divide mitotically to produce more oogonia, all of which are diploid. By the third month of embryonic development, the oogonia begin to develop into **primary oocytes** by starting meiosis. However, meiosis is not completed; it is arrested in the prophase of meiosis I until ovulation.

    b. **Menarche.** The beginning of ovulation at the onset of puberty is called **menarche.** A woman does not possess any fertile eggs until this time.

    (1) At birth, a girl infant has approximately 400,000 primary oocytes, all of which are awaiting completion of meiosis I.

    (2) Only a fraction of these primary oocytes survive until puberty, and even fewer (approximately 400) are ovulated.

    c. **Ovulation.** The primary oocytes that are ovulated resume meiosis I, forming the **secondary oocyte** and the **first polar body.** At ovulation, the secondary oocyte is released from the ovary into the abdominal cavity and is then picked up by **fimbria** (cilia lining the oviduct), which draw the oocyte inside the oviduct.

    d. **Fertilization.** Completion of meiosis II occurs only if the secondary oocyte becomes fertilized. The products of meiosis II are the mature **ovum** and the **second polar body.** By the time the ovum is mature, it is already fertilized and is technically a **diploid zygote.**

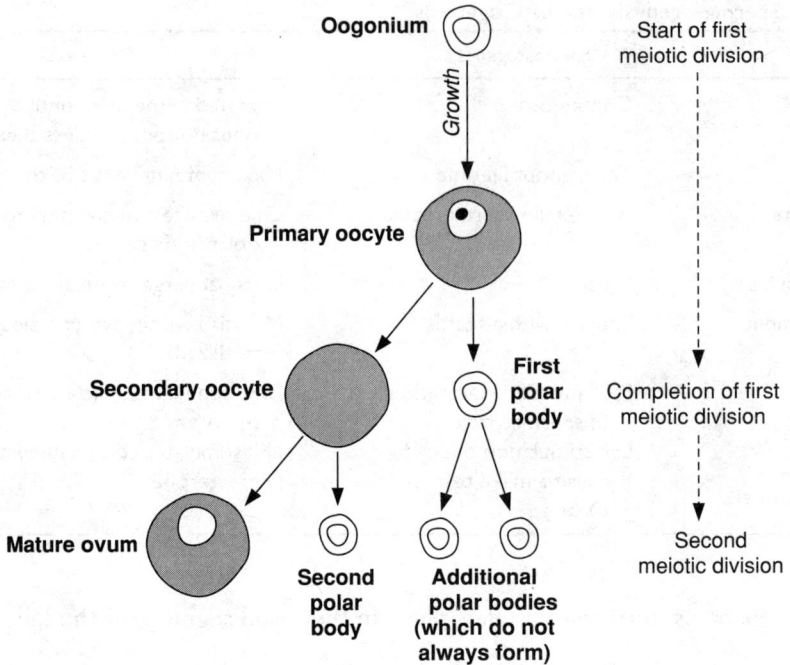

Figure 17-5. Oogenesis.

2. **Division of cytoplasm.** Unequal division of cytoplasm occurs during oogenesis. Most of the cytoplasm of an oogonia is ultimately transferred to only one of its meiotic products, the mature ovum. The remainder of the meiotic products, the polar bodies, contain almost nothing but a nucleus. This **unequal cytoplasmic division** ensures that the mature egg has enough cytoplasm and stored proteins to allow the zygote to begin development and to survive.

3. **Hormonal control.** The ovaries and the pituitary have a reciprocal effect on one another (Table 17-3). For a review of the endocrine glands, the hormones they produce, and the activities they control, see Chapter 9.

   a. During oogenesis, cells surrounding the developing ovum **assist in the maturation process** and form a spherical structure called a **follicle.**

   (1) The follicle continues to grow under the influence of **FSH** released from the anterior pituitary gland. In addition, FSH **stimulates the cells of the follicle to produce estrogen.**

   (2) As **estrogen levels increase,** the follicles begin to inhibit the further release of FSH.

   b. **Ovulation** is achieved with the aid of **LH.** After ovulation, the cells of the follicle are transformed into a structure called the **corpus luteum.** LH stimulates the corpus luteum to **produce progesterone.**

C. **Menstrual Cycle**

The menstrual cycle is approximately 28 days. It occurs from menarche to menopause, and it can be divided into four phases (Figure 17-6).

TABLE 17-3. Female Reproductive Hormones

Endocrine Gland and Hormones	Site of Action	Activities Controlled
Anterior Pituitary		
Follicle-stimulating hormone (FSH)	Ovary	Stimulates development of follicles; with LH, stimulates ovulation and secretion of estrogen
Luteinizing hormone (LH)	Ovary	Stimulates ovulation and development of corpus luteum (including progesterone release)
Prolactin	Breast	Milk production
Ovary		
Estrogens	General	Puberty: growth spurt; development of secondary sex characteristics (e.g., pubic and axillary hair, breast development)
		Adult: maintenance of secondary sex characteristics; oogenesis
	Reproductive structures	Monthly preparation of endometrium; makes cervical mucus more alkaline and thinner
Progesterone	Uterus	Maintains endometrium for pregnancy
	Breasts	Stimulates development of mammary glands

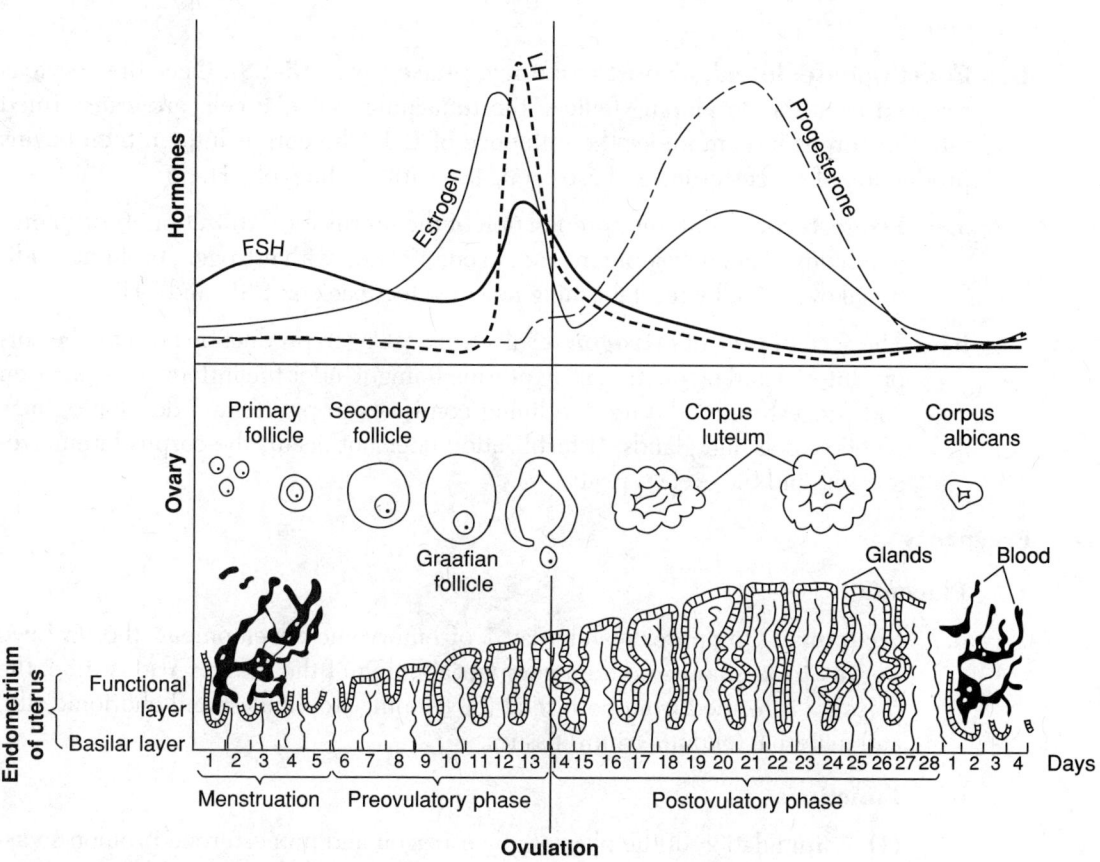

Figure 17-6. The menstrual cycle. (Reproduced with permission from Ville CA, Solomon EP, Davis PW: *Biology.* Philadelphia, Saunders College Publishing, 1985, p 939.)

1. **First phase: menstruation (days 1–5).** Maintenance of the uterine lining depends on the continued presence of progesterone. Initially, the progesterone is produced by the corpus luteum under the stimulation of LH. However, the life span of the corpus luteum is approximately 10 days, and if implantation of a fertilized egg does not occur, the corpus luteum regresses. This event coincides with a decrease in progesterone concentration, which induces the lining of the uterus to slough and causes the characteristic bleeding.

2. **Second phase: follicular, or preovulatory, phase (days 6–13).** A maturing oocyte is surrounded by a growing mass of follicular cells that release estrogen in response to FSH stimulation. The estrogen helps to prepare the uterine lining for fertilization. Normally, FSH and LH are suppressed by high levels of estrogen and progesterone. However, during menstruation, both of these hormones are abruptly withdrawn, removing the negative feedback inhibition and causing a rise in FSH and LH at the beginning of the follicular phase. As the follicular cells begin to produce estrogen, the negative feedback is reestablished, causing a decline in FSH and LH levels toward the end of the follicular phase.

3. **Third phase: ovulation (day 14).** The small, abrupt rise in estrogen near the end of the follicular phase causes a surge in FSH and LH release. High plasma concentrations of estrogen appear either to suspend negative feedback or to induce a positive feedback on FSH and LH. Regardless of the mechanism, this FSH/LH surge causes the release of the developing oocyte from the follicle, which is the process of ovulation.

4. **Fourth phase: luteal, or postovulatory, phase (days 15–28).** Once the oocyte is released from the developing follicle, the remaining follicular cells are transformed into the corpus luteum under the influence of LH. The corpus luteum then begins production of progesterone and estrogen, also with the help of LH.

   a. **Progesterone** continues preparation of the uterus for fertilization. It promotes mammary gland development and, in conjunction with estrogen, prohibits additional ovulation by reestablishing negative feedback on FSH and LH.

   b. The second peak of **estrogen** results from estrogen production both by the corpus luteum and by the maturing uterine lining. Under the influence of estrogen and progesterone, the uterine lining continues to proliferate, developing new blood vessels and glands. If fertilization does not occur, the corpus luteum regresses, and the cycle repeats.

D. **Pregnancy**

   1. **Placenta**

      a. **Structure.** On approximately day 7 of embryonic development, the **embryo begins to implant** into the endometrial lining of the uterus. At this time, the placenta is formed from both fetal (developing trophoblast and chorionic villi) and maternal (endometrium) tissues.

      b. **Functions**

         (1) Stimulation of the placenta by estrogen and progesterone promotes **vascularization, glandular development, secretory activity, accumulation of fluid, and endometrial proliferation.**

(2) The placenta is the **organ of exchange between the fetus and the mother.** The placenta provides:

(a) Nutrients and oxygen to the fetus

(b) Removal of waste products from the fetal circulation

(c) An endocrine organ for the production of estrogen, progesterone, and human chorionic gonadotropin (hCG)

2. **Human chorionic gonadotropin.** hCG plays a key role in pregnancy because it maintains the corpus luteum beyond its normal 10-day life span. As a result, progesterone production continues, preventing sloughing of the endometrium and subsequent ovulation, both of which would be detrimental to the developing fetus. By approximately the second month of pregnancy, the placenta is able to produce progesterone in sufficient quantities so that a positive feedback loop is established, ensuring proper maintenance of the placenta throughout gestation. Excess hCG is excreted in the mother's urine and is the basis for most pregnancy tests.

3. **Fetal membranes.** There are four membranes that surround the fetus. However, only two of the membranes contribute to fetal structures. They are the yolk sac and the allantois (Figure 17-7).

   a. The **amnion** is the closest membrane to the fetus. It creates a fluid-filled cavity (i.e., the amnionic cavity) around the developing embryo. The fluid is derived from the maternal circulation and from fetal excretory products. The amnionic cavity serves several functions:

   (1) Absorbs shocks

   (2) Allows fetal movement and growth

   (3) Prevents adhesion of the amnion to the fetus

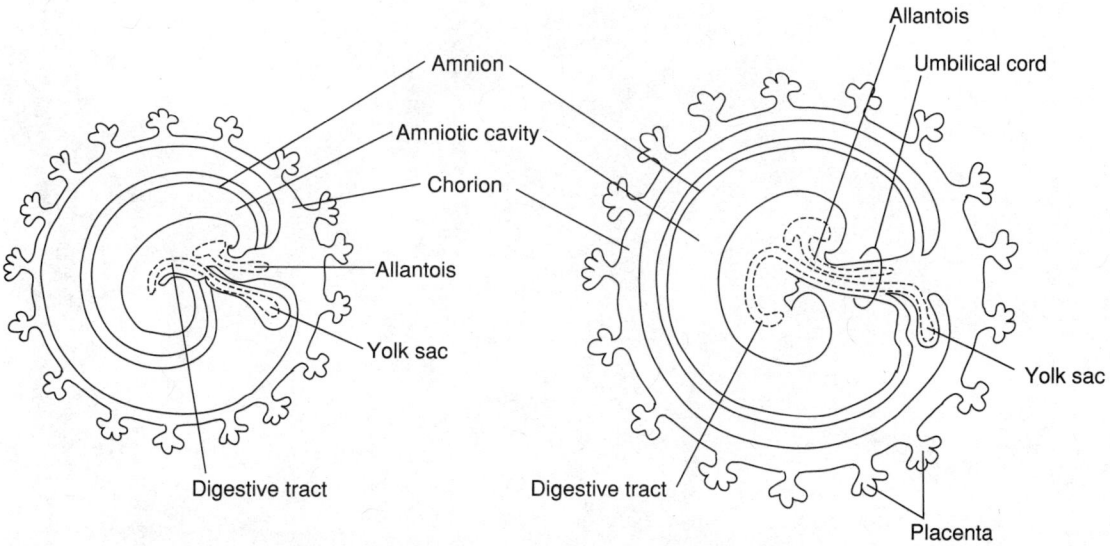

Figure 17-7. Structure of the fetal membranes. (Reproduced with permission from Ville CA, Solomon EP, Davis PW: *Biology.* Philadelphia, Saunders College Publishing, 1985, p 965.)

- (4) Regulates temperature
- (5) Protects the fetus during birth

b. The **chorion** is the fetal contribution to the placenta.

c. The **yolk sac** connects to the umbilical cord within the chorionic cavity. Functions include:
- (1) Initial blood cell development
- (2) Formation of the germ cells
- (3) Nutrient transfer before formation of a functional placenta
- (4) Provision of endodermal derivatives of the gastrointestinal tract and respiratory system

d. The **allantois** contributes to the following:
- (1) Blood cell formation
- (2) Connection between the bladder and umbilical cord
- (3) Umbilical artery and vein formation

Development 18

I. Human Embryology

A. **Fertilization,** which is the **union of a spermatozoon and an ovum,** marks the beginning of human embryonic development.

1. **Functions.** Fertilization serves three functions:

    a. Restores the diploid state when the haploid nuclei of the sperm and ovum unite

    b. Determines the sex of the offspring, which is determined by the sex chromosome carried by the sperm cell

    c. Initiates development

2. **Capacitation and the acrosome reaction**

    a. **Capacitation.** Upon entry into the female reproductive tract, spermatozoa are not fully capable of fertilization. The spermatozoa must first undergo capacitation, which is a process that **strips the coat of glycoprotein** molecules off the surface of the spermatozoa. (The glycoproteins adsorb to the spermatozoa during maturation in the epididymis.) These molecules are removed by the proteolytic enzymes and high ionic strength of the estrogen-primed uterus.

    b. **Acrosome reaction.** After capacitation, spermatozoa are capable of activation via the acrosome reaction.

        (1) The acrosome is the **cap-like structure** at the head end of the sperm.

        (2) During the acrosome reaction, this cap breaks down, releasing enzymes (e.g., hyaluronidase) that **degrade the cells of the corona radiata** and help the spermatozoon to penetrate the oocyte (Figure 18-1).

        (3) It is important that activation occurs in close proximity to the oocyte because once spermatozoa are activated, their viability is greatly reduced.

3. **Fusion of membranes and pronuclei**

    a. **Fusion.** Once a spermatozoon has penetrated the corona radiata, it comes into contact with the **zona pellucida,** which is a protective membrane that surrounds the oocyte and stimulates the zona reaction. After passing through the zona pellucida, **the plasma membrane of the spermatozoon fuses with that of the oocyte** (see Figure 18-1).

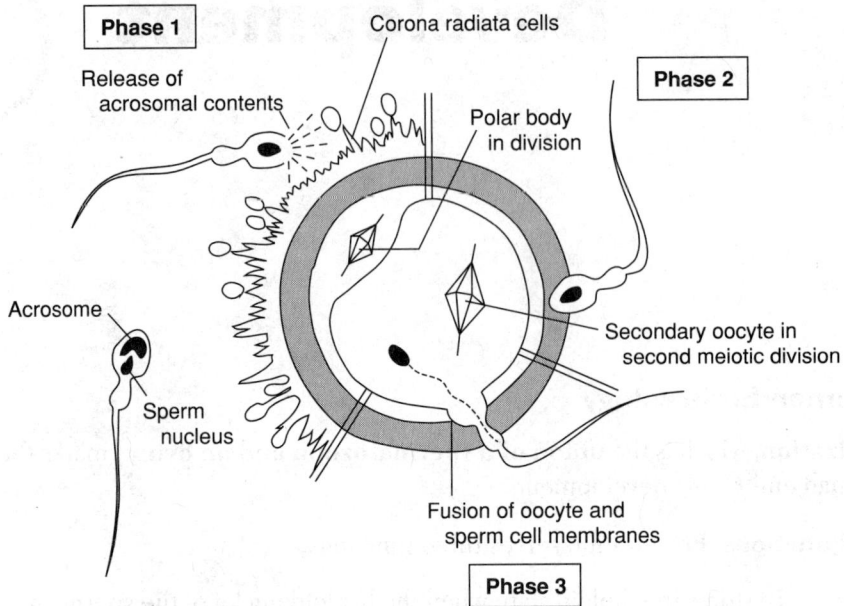

Figure 18-1. Phases of oocyte penetration.

  b. **Male and female pronuclei.** Fusion stimulates the completion of meiosis II (see Chapter 17) and allows for entry and formation of the **male pronucleus** in the cytoplasm of the newly formed zygote (Figure 18-2). When the male pronucleus and **female pronucleus** come into contact, their nuclear membranes break down, their chromosomes intermingle, and a diploid nucleus is established. This completes the process of fertilization.

4. **Advantages of large numbers of spermatozoa.** Movement of spermatozoa is undirected, so large numbers help to ensure that a few come into contact with the oocyte. The acidic pH of the uterus and phagocytic degradation greatly decrease the number of spermatozoa. Multiple spermatozoa can work together in a coordinated fashion to break through the corona radiata.

5. **Barriers to polyspermy.** Fusion of one oocyte with one spermatozoon results in creation of a genetic diploid state. However, if multiple spermatozoa entered the oocyte,

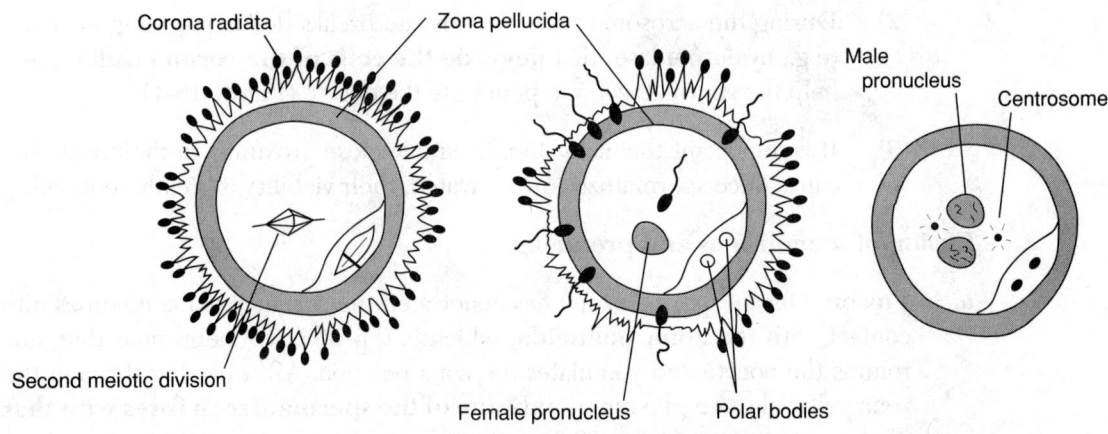

Figure 18-2. Formation of male and female pronuclei.

polyploidy could arise. Polyspermy results in excess chromosomes and is incompatible with life. There are several barriers to multiple spermatozoa entry.

    a. **Negative charge.** Spermatozoon penetration stimulates **hyperpolarization of the membrane of the zygote.** This enhanced negative charge repels further binding of spermatozoa.

    b. **Cortical reaction.** Hyperpolarization mobilizes calcium, the displacement of which results in **fusion of cortical granules** in the cytoplasm of the zygote with the plasma membrane.

    c. **Zona reaction.** The cortical reaction **releases enzymes that modify the zona pellucida,** which impairs subsequent spermatozoa penetration. Strongly acidic glycoproteins are released along with the modifying enzymes. These proteins polymerize on the surface of the zona pellucida and form a protective coat around the zygote.

B. Cleavage is a rapid series of mitotic divisions. Upon fertilization, there is a dramatic increase in the metabolic rate, oxygen consumption, and protein synthesis of the zygote, all of which prepare the zygote for cleavage. During cleavage, the embryo migrates down the fallopian tube.

1. Cleavage **increases the number of cells** comprising the embryo but without any protoplasmic growth because each interphase period is too brief to allow the cells to grow. Hence, with each division, the **cells become progressively smaller.** The small size of the cells enables them to move about with ease, partitioning the zygote into distinct regions in preparation for future developmental events.

2. The zygote is called a **morula** when it reaches the 16-cell stage. This state is reached as the embryo enters the uterus. At this time, the zona pellucida is fully dissolved.

    a. The morula consists of a group of centrally located cells, the **inner cell mass,** which gives rise to the embryo proper.

    b. The morula contains a surrounding outer cell layer, the **outer cell mass,** which later forms the **trophoblast.** The trophoblast is the fetal contribution to the placenta (see Chapter 17).

C. Blastulation occurs when fluid begins to accumulate between the inner and outer cell masses, creating an internal cavity called the **blastocoele,** or **blastocyst cavity** (Figure 18-3A).

1. The **embryoblast** is the inner cell mass that polarizes to one end of the embryo.

2. At this time, the developing zygote as a whole is referred to as **the blastocyst,** and implantation begins.

3. The cells of the trophoblast invade the endothelial layer of the uterus, establishing the initial connections of the placenta (see Figure 18-3B).

D. Gastrulation occurs during the third week of embryonic development and creates a **trilaminar disk** composed of the three germ layers: **endoderm, mesoderm,** and **ectoderm.**

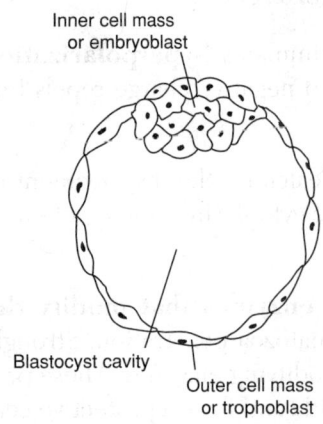
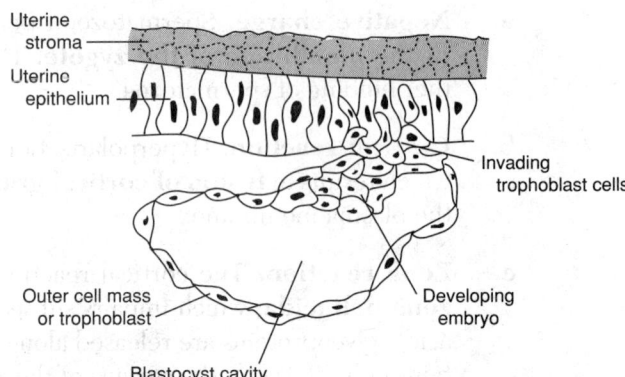

Figure 18-3. Blastocyst formation.

1. Before gastrulation, cells of the embryoblast migrate to form two parallel planes of cells—the **hypoblast** and the **epiblast** (Figure 18-4). When the **bilaminar disk** is formed, the amnionic cavity and the primitive yolk sac are formed coordinately (see Chapter 17).

2. Gastrulation begins with an **invagination of the epiblast** along its superior surface, which produces the **primitive streak** (Figure 18-5).

    a. **Invagination.** As the epiblast invaginates, some of its cells differentiate and then migrate between the hypoblast and the epiblast.

    b. **Formation of the trilaminar disk.** Some of these differentiated cells displace the hypoblast to form the embryonic endoderm, and others come to lie between the newly formed endoderm and the epiblast to form the embryonic mesoderm.

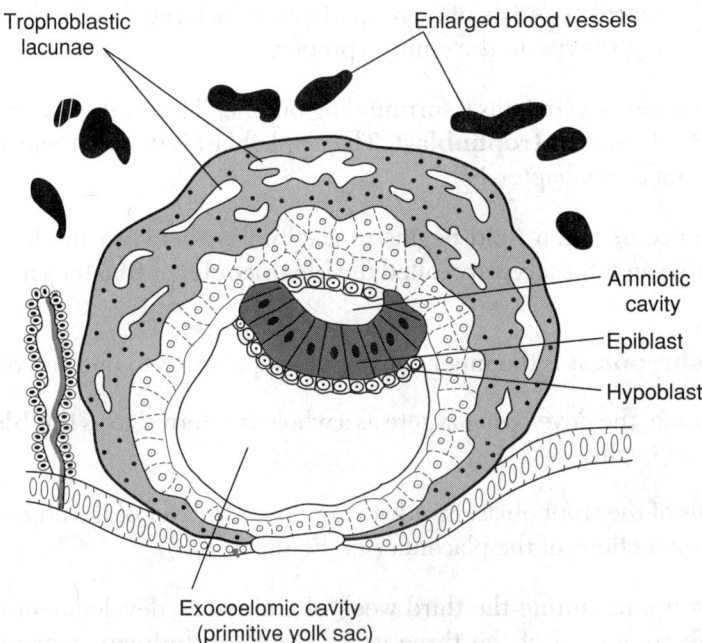

Figure 18-4. The bilaminar disk.

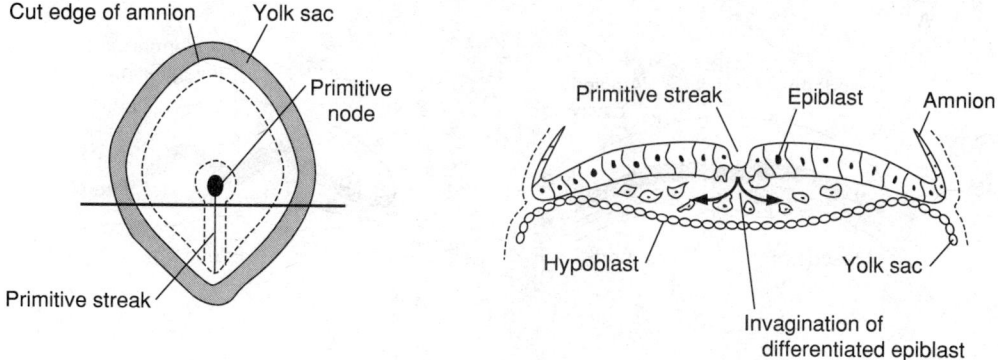

Figure 18-5. Formation of the trilaminar disk.

The remaining undifferentiated epiblast is referred to as the ectoderm, and the trilaminar disk is completed.

    c. **Differentiation.** The three germ layers of the trilaminar disk eventually give rise to the anatomic features of the body (Table 18-1).

E. **Neurulation** refers to neural plate formation and the development of a neural tube. The **brain and spinal cord,** which arise from these embryonic structures, are among the first organs to develop.

    1. The **notochord.** During the second week of embryonic development, mesodermal cells differentiate and form a **cylindrical rod along the length of the embryo** (the notochord). This serves as a flexible skeletal axis for all chordates. In vertebrates, the notochord is later replaced by a bony vertebral column, and its remnants become part of the intervertebral disks.

    2. The **neural plate.** Cells of the notochord secrete factors that induce the overlying ectoderm to thicken and form the neural plate (Figure 18-6). The neural plate invaginates to become the **neural groove** and ultimately the **neural tube,** which later forms the brain and the spinal cord.

TABLE 18-1. Germ Layer Derivatives.

Endoderm	Mesoderm	Ectoderm
Linings of internal organs and ducts (respiratory, gastrointestinal, urinary, eustachian tube, tympanic cavity)	Cartilage and bone	Central and peripheral nervous system
	Muscle (skeletal, cardiac, smooth)	Epidermis and associated structures (e.g., hair, nails)
	Dermis and connective tissue	

Parenchyma of:
  Liver
  Tonsil
  Thyroid
  Thymus
  Pancreas
  Parathyroid
	Blood cells	Cornea and lens of eye

	Cardiovascular, excretory, reproductive, and lymphatic systems	Inner ear
		Tooth enamel

	Organs:
	  Adrenal cortex
	  Gonads and their ducts
	  Spleen
	  Kidneys
		Nasal olfactory epithelium

		Mammary and pituitary glands
		Adrenal medulla

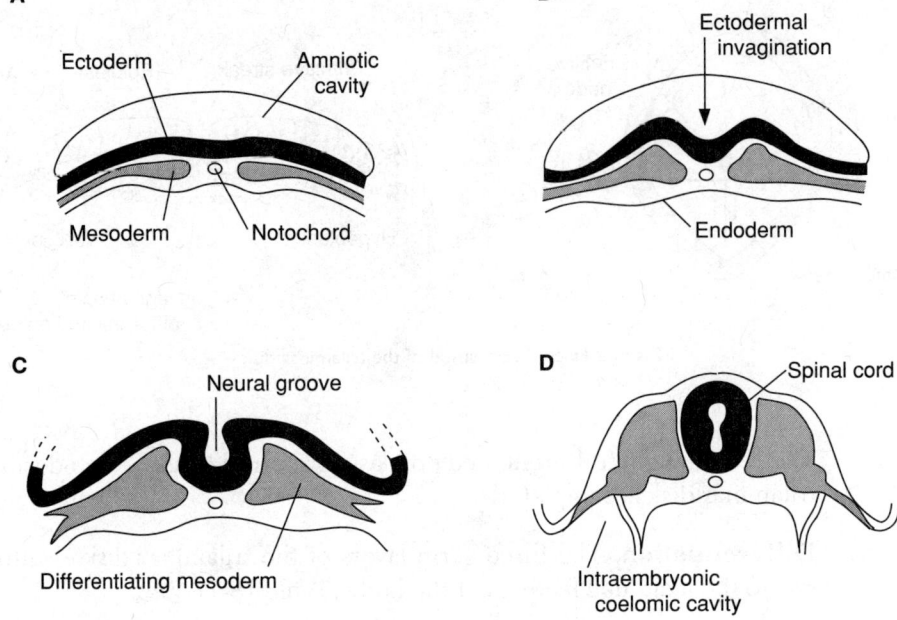

Figure 18-6. Neural tube formation.

II. Developmental Mechanisms

Embryonic development is a highly complex sequence of events that must be exquisitely orchestrated. A high degree of communication must take place between cells to allow for tissue, organ, and system development. Some general developmental concepts are discussed in the following paragraphs.

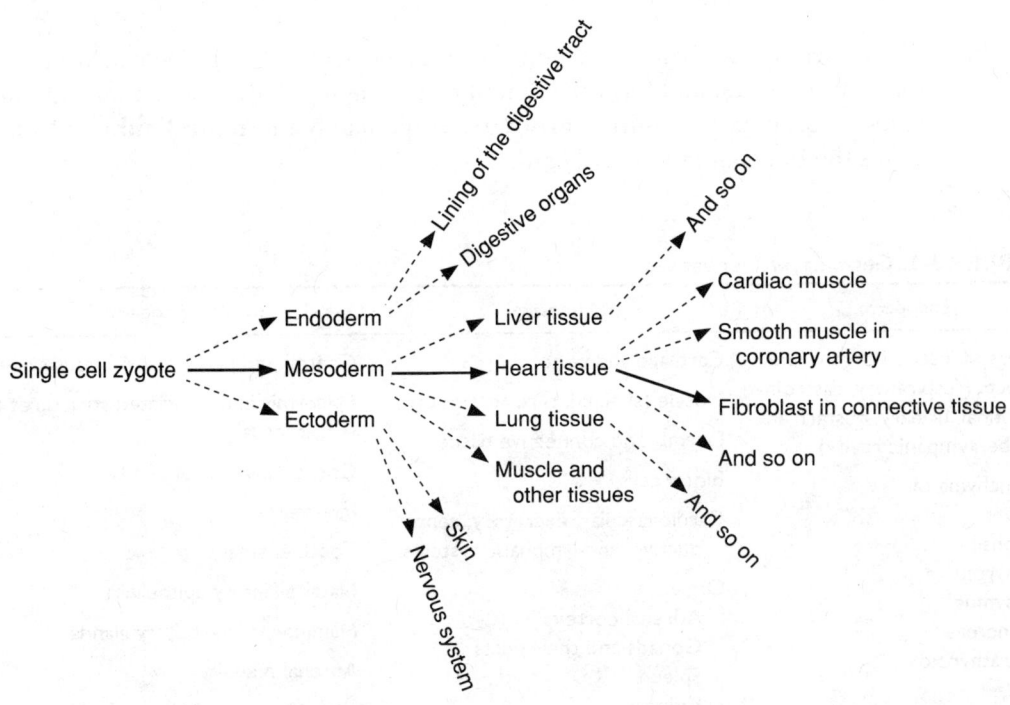

Figure 18-7. The process of determination.

A. Differentiation is the process by which immature cells or tissues develop into mature cells or tissues with specialized functions. Differentiation is accompanied by specific protein expression and cytoskeletal modifications so that structures are created that assist the functions required of the specialized, mature cells.

B. Determination is the process that occurs as cells undergo development and their destiny becomes progressively more restricted. For example, a single-cell zygote has the potential to develop into any type of specialized eukaryotic cell (Figure 18-7). As the three germ layers are formed, the destiny of a cell within a given layer becomes limited, although many possibilities are still possible. The cell continues to differentiate until a single pathway of differentiation is chosen.

C. Induction is the process by which a chemical mediator released from one part of the embryo causes a specific morphogenic effect in another part by inducing a particular developmental pathway. For instance, the dermis of the skin, which is derived from the mesoderm, releases a substance that causes the overlying ectoderm to differentiate into epidermis. As a consequence of simultaneous induction of neighboring cells, tissues and ultimately organs can be formed.

SECTION III

Genetics, Evolution, and Botany

Genetics 19

I. Introduction

A. Basic Definitions

1. **Genetics** is the study of the heritable information in organisms.

2. A **gene** is a unit of genetic information encoded in DNA (see Chapter 4).

   a. The **genotype** of an organism is its genetic makeup, including the information in all its genes. The **potential** characteristics or traits of an organism are encoded in its genotype.

   b. The **phenotype** of an organism is its actual or **expressed** traits. A specific phenotype results from the expression of specific genes.

3. Genes are carried on **chromosomes,** which are discrete structures made up of DNA and associated histone proteins. The position of a gene on a chromosome is called its **locus.**

4. A **diploid** cell contains two sets of chromosomes, one set inherited from each parent. Each chromosome is paired with another chromosome. The two chromosomes of a pair are called **homologous chromosomes** or **homologs.**

   a. Homologous chromosomes carry genes for the same traits at corresponding loci. Therefore, homologs are functionally similar. However, homologs do not necessarily contain identical genetic information.

   b. If N is the number of functionally different chromosomes, then diploid cells have $2N$ chromosomes. Each species has a characteristic diploid number of chromosomes: **Humans have 46** ($2N = 46$), dogs have 78, fruit flies have 8.

   c. Homologs look alike in the **karyotype** of an organism, which is a stained preparation of all the chromosomes. In a karyotype, the different types of chromosomes can be distinguished by size, centromere location, and pattern of dark and light bands.

5. The **somatic cells** are the diploid cells that make up an organism. All the somatic cells in an organism have the same genotype.

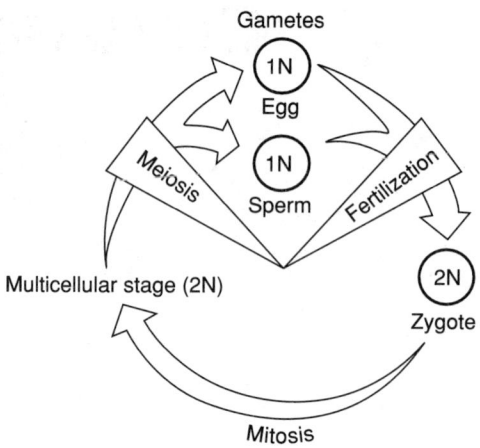

Figure 19-1. Sexual reproduction.

- **B. Sexual Reproduction** (Figure 19-1). **Gametes (germ cells)** are produced by diploid organisms. A germ cell, which is **haploid,** has half of the diploid number of chromosomes. **Human germ cells have 23 chromosomes** ($2N = 46$; $1N = 23$), which is one of each type of chromosome.

    1. The **germ line** of an organism includes the gametes and the cells that give rise to them. In the germ line, new gametes are produced through a special cell division called **meiosis.**

    2. In sexual reproduction, **each parent contributes one set of homologs to the offspring.** Sexual reproduction results in unique combinations of genetic information in each offspring.

        a. In **fertilization,** the father's gamete (spermatozoon) fuses with the mother's gamete (egg) to produce the diploid **zygote** or fertilized egg.

        b. The zygote grows by **mitotic divisions.**

II. Mendel and the Principles of Heredity

- **A. Mendelian Inheritance**

    Gregor Mendel first proposed many of the fundamental principles of genetics. In the late nineteenth century, he recognized the existence of units of heritable information for the observable traits of organisms. This was the first concept of the gene.

    1. Mendel studied the **inheritance of traits in pea plants.** In a single plant, each trait occurs in either of two forms, for example, yellow or green seeds, round or wrinkled seeds, short or tall stems. The following abbreviations are used in Mendelian genetics.

        a. **P—parental generation.** Each parental strain is pure breeding, meaning that it produces offspring with only one of the alternative forms of a trait.

        b. **F1—first offspring (filial) generation.** These plants are crossed with each other to produce the F2 generation.

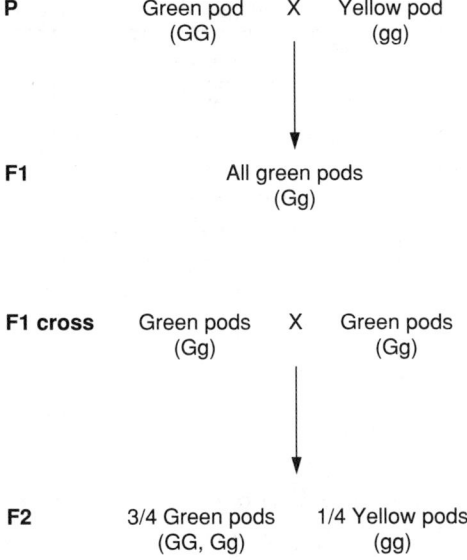

Figure 19-2. Mendel's cross of two pure-breeding strains with contrasting forms of a trait (i.e., color). P = parental generation; F1 = first offspring (filial) generation; F2 = second offspring generation, produced by crossing two F1 plants.

    c. **F2—second offspring generation.** These plants are produced by crossing two F1 plants.

2. **Mendel's method** was to cross two pure-breeding strains with contrasting forms of a trait (Figure 19-2).

3. **Mendel's conclusions** include the following.

    a. The information for a trait (pod color) is a discrete, heritable factor (gene).

    b. Genes come in pairs, called **alleles.** Alleles are the different forms of a gene for a particular trait.

    c. In the diploid organism, a **dominant allele** of a gene may mask the phenotypic expression of the **recessive allele.** For the pod color gene, G is the dominant allele (green); g is the recessive allele (yellow).

        (1) The pure-breeding **P-generation plants are homozygous** for their alleles: GG (green) and gg (yellow).

        (2) All the **F1 plants are hybrids,** or organisms carrying both alleles of a particular gene. The F1 plants are **heterozygous** for the two alleles. Although Gg is the F1 genotype, only the dominant allele G shows in the phenotype (green).

    d. The two alleles segregate when a hybrid produces gametes. This **segregation** occurs randomly, so that half of the gametes receive the G allele, and half receive the g allele. This is **Mendel's first law—the law of random segregation.**

    e. The F1 cross described above is called a **monohybrid cross** (one trait is segregating). In fertilization, each F1 parent contributes a gamete containing either

TABLE 19-1. Prediction of F2 Phenotypes Using a Punnett Square; Monohybrid Cross

		F1 Spermatozoa	
		½ G	½ g
F1 Eggs	½ G	¼ GG	¼ Gg
	½ g	¼ gG	¼ gg

F1: Gg (green) × Gg
F2 genotypes: ¼ GG, 2/4 Gg, ¼ gg (1:2:1)
F2 phenotypes: ¾ green, ¼ yellow (3:1)

G or g to the offspring. Which gametes are combined is determined by a random event.

4. The **F2 genotypes can be predicted** using the **Punnett square method** (Table 19-1). The alleles contributed by the spermatozoon and egg are placed along the top and left side, respectively. The F2 offspring resulting from each fertilization event are placed inside the squares.

5. The **genotype** of the phenotypic green plant **can be determined by a testcross.**

   a. If the genotype is GG, all the progeny will be Gg (green).

   b. If the genotype is Gg, half the progeny will be Gg (green) and half will be gg (yellow).

6. Segregation and fertilization are both random events. Therefore, the **rules of probability** can be used to predict genotypes or phenotypes in mendelian crosses.

   a. The **multiplication rule** states that the probability that two events will both happen equals the product of the probabilities that each independent event will happen. For example, in the F1 cross (Gg × Gg), the probability that an F2 plant will have yellow pods can be determined as follows:

      (1) A plant with yellow pods (gg) must receive a g allele from both the egg and the spermatozoon. The probability of g in the egg = ½. The probability of g in the spermatozoon = ½.

      (2) The probability of g in the egg and in the spermatozoon is the product of the probabilities of each: ½ × ½ = ¼. Therefore, 25% of the F2 plants should have yellow pods.

   b. The **addition rule** states that the probability that either of two events will happen is predicted by adding the probabilities of the alternate events. For example, in the F1 cross, the probability that an F2 plant is Gg can be determined as follows:

      (1) The probability of G from the egg and g from the spermatozoon = ½ × ½ = ¼.

      (2) The probability of g from the egg and G from the spermatozoon = ½ × ½ = ¼.

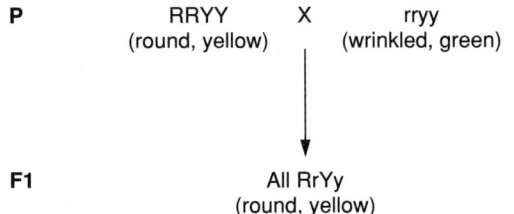

Figure 19-3. Mendel's dihybrid cross in which two different traits (i.e., color and shape) are segregating. Y = yellow; y = green; R = round; r = wrinkled.

(3) The probability of either of these events happening is the sum of the probabilities of each: ¼ + ¼ = ½. Therefore, 50% of the F2 plants should be Gg.

B. Independent Assortment

Mendel's second law is the **law of independent assortment.** This law states that the genes for different traits are inherited (assorted) independently from each other. Mendel performed a **dihybrid cross** in which two different traits segregated (Figure 19-3).

1. The phenotypes of the F2 progeny fit the Punnett square for independent inheritance of the genes for seed color and shape (Table 19-2).

2. As in the monohybrid cross, each of the traits exhibits a 3:1 phenotypic ratio (i.e., 12 round : 4 wrinkled).

III. Population Genetics

A. Population genetics is concerned with the relative **frequencies of dominant and recessive alleles** in a population of interbreeding organisms.

1. The **gene pool** is the sum of all the genotypes in a population at any given time.

2. A population is in **genetic equilibrium** when the gene pool remains constant from one generation to the next. Genetic equilibrium occurs only under the following "ideal" conditions:

   a. Mating is completely random.

   b. The population is very large.

TABLE 19.2. Prediction of F2 Phenotypes in a Dihybrid Cross

		F1 Spermatozoa			
		RY	Ry	rY	ry
F1 Eggs	RY	RRYY	RRYy	RrYY	RrYy
	Ry	RRYy	RRyy	RrYy	Rryy
	rY	rRYY	rRYy	rrYY	rrYy
	ry	rRyY	rRyy	rryY	rryy

F2 phenotypes: 9/16 round yellow, 3/16 round green, 3/16 wrinkled yellow, 1/16 wrinkled green (9:3:3:1)

- c. There is no net change in the gene pool caused by mutations.
- d. The population is isolated (i.e., there is no migration in or out).
- e. All genotypes have equal reproductive success.

B. The **Hardy-Weinberg principle** states that, under conditions of genetic equilibrium, the frequencies of alleles in a gene pool remain constant from generation to generation. The frequency of an allele in a population determines the proportion of gametes that will contain that allele. The random probability of combinations of alleles results in predictable genotype frequencies in the offspring.

1. **Hardy-Weinberg equation.** The **genotype frequencies in a gene pool** can be calculated if the allele frequencies are known, and vice versa.
   - a. For the two alleles A and a, let p represent the frequency of the dominant allele A, and let q represent the frequency of the recessive allele a.
   - b. The **sum of the allele frequencies** must equal 100% of the genes for that locus in the population: $p + q = 100\%$, or $p + q = 1$. If the frequency of one allele is known, the other can be derived by these equations: $1 - p = q$ or $1 - q = p$.
   - c. The **frequencies of the genotypes** in the offspring can be predicted from the allele frequencies and from the probability of each combination of alleles.
     - (1) The combination of two gametes containing the same allele occurs with a probability of:
       - (a) $p \times p = p^2$, for the homozygous dominant genotype (AA)
       - (b) $q \times q = q^2$, for the homozygous recessive genotype (aa)
     - (2) The combination of gametes containing different alleles occurs with the probability of: $(p \times q) + (p \times q) = 2pq$, because the heterozygote (Aa) can be formed in either of two ways.
   - d. The **sum of the genotype frequencies must equal 100% (or 1):**
     - (1) $p^2 + 2pq + q^2 = 1$
     - (2) AA + Aa + aa = 1
   - e. The **allele frequencies** in this generation **can be determined from the genotype frequencies.**
     - (1) All the gametes from the homozygous AA and half the gametes from the heterozygous Aa will contain the p allele. Therefore, p = frequency AA + ½ frequency Aa.
     - (2) Likewise, q = frequency aa + ½ frequency Aa

2. **A Hardy-Weinberg problem**
   - a. It is given that p = the frequency of allele A = 0.7, and q = the frequency of allele a = 0.3.

b. From the equation $p^2 + 2pq + q^2 = 1$, the **genotype frequencies** in the next generation are as follows:

  (1) AA = 0.7 × 0.7 = 0.49, or 49%.

  (2) aa = 0.3 × 0.3 = 0.09, or 9%.

  (3) Aa = 2(0.7 × 0.3) = 0.42, or 42%

c. The **allele frequencies** in this population are as follows:

  (1) A = 0.49 + ½(0.42) = 0.7

  (2) a = 0.09 + ½(0.42) = 0.3

d. Because the allele frequencies remained constant, this population is said to be in **Hardy-Weinberg equilibrium.**

3. **Application.** The Hardy-Weinberg equation applies to **nonevolving populations.** If the gene pool of a population is changing, then the actual frequencies in the population will not fit those predicted by the equation.

IV. Meiosis

A. The **process of meiosis** (Figure 19-4). A diploid (2N) organism produces haploid (1N) gametes through meiosis, which is a special division that **reduces the chromosome number by half.** Meiosis is necessary for fusion of two gametes to produce a zygote with a diploid number of chromosomes.

1. Meiosis involves **two sequential divisions**—meiosis I and meiosis II. The entire process results in four different haploid daughter cells, which form the gametes (Table 19-3).

2. **Meiosis I** is the reductional division in which the chromosome number is halved. In meiosis I, homologous chromosomes are paired, then separated.

   a. **Prophase I.** The first phase takes approximately 90% of the time for meiosis.

      (1) Homologous chromosomes pair through a process called **synapsis.** The genes on one homolog line up side-by-side with the same genes on the other homolog. A complex of proteins forms between the paired chromosomes. Four **chromatids,** the two sisters of each homolog, are involved at this stage.

      (2) The paired homologs undergo **crossing over** at several sites. In this process, chromatids break, and homologous segments are exchanged between the paired chromosomes. The visible structures formed by crossovers are called **chiasmata.** Chromosome pairs are physically held together by their chiasmata through the next stage of meiosis.

      (3) The **meiotic spindle** also develops during prophase I.

   b. **Metaphase I.** At this stage, the homologous pairs of chromosomes line up at the metaphase plate. The sister chromatids of each chromosome are attached to the

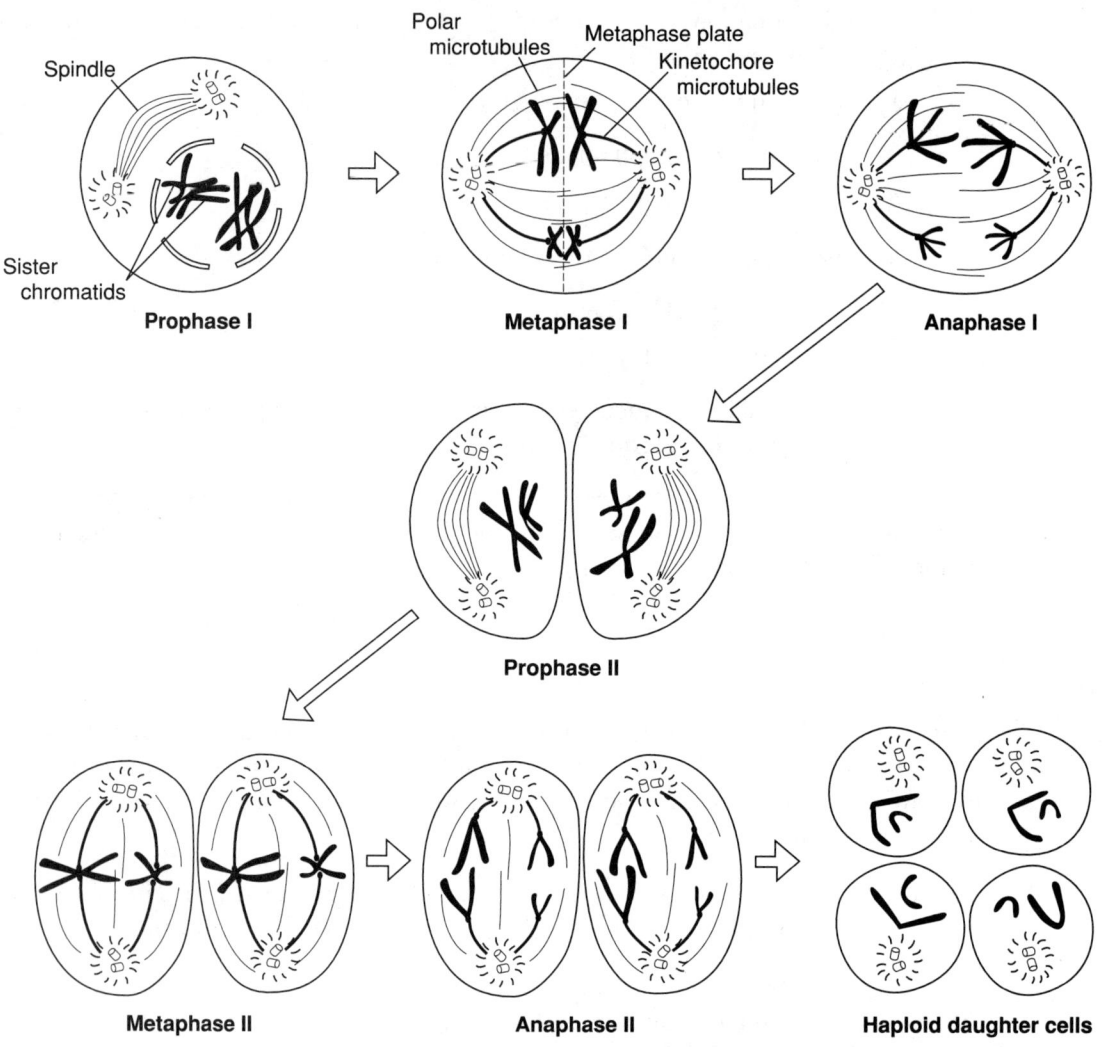

Figure 19-4. The process of meiosis.

TABLE 19-3. Summary of Meiosis

Stage	Event
Interphase I	Chromosome replication; sister chromatids attach at centromere
Prophase I	Synapsis of homologous pairs, crossing over at chiasmata; spindle forms
Metaphase I	Homologous pairs line up at metaphase plate, with centromeres attached to microtubule fibers from opposite poles
Anaphase I	Homologous pairs of chromosomes separate and move toward opposite poles; centromeres do not split
Metaphase II	Individual chromosomes, consisting of two chromatids, line up at metaphase plate
Anaphase II	Centromeres split; sister chromatids separate and move toward opposite poles

same pole. The paired homologs are attached to opposite poles and are oriented randomly.

    c. **Anaphase I.** The crossovers are resolved. The homologs move, centomeres first, toward opposite poles. (The sister chromatids of each chromosome remain together.)

    d. **Telophase I–cytokinesis–interphase II.** The cell divides into two cells, each of which contains a haploid chromosome number. Interphase II is of very short duration, and no DNA replication occurs.

3. **Meiosis II** is called the equational division. In meiosis II, sister chromatids are separated from each other (as in mitosis).

    a. During **prophase II,** the spindle develops.

    b. In **metaphase II,** the chromosomes line up on the metaphase plate, with sister chromatids facing opposite poles.

    c. In **anaphase II–telophase II,** the sister chromatids are separated and move as individual chromosomes toward the poles. Four daughter cells are produced, each with a haploid number of chromosomes.

B. **Comparison of Meiosis I and Mitosis** (Table 19-4). Mitosis produces two daughter cells that are genetically identical to the parent cell and to each other. Meiosis produces four haploid daughter cells that differ genetically from the parent cell and from each other.

C. **The Role of Meiosis in Genetic Variability**

Meiosis and fertilization involve chromosome sorting and recombining, resulting in new combinations of genes in the offspring. The genes in a population are reshuffled with each successive generation.

1. In meiosis I, the homologs orient randomly at the equator. This results in a **random segregation** to one pole or the other and an equal probability for either homolog to be present in a particular gamete.

    a. Each of the homologous pairs (23 in humans) assorts independently of the others. This means that **a human gamete has a possibility of 2^{23}, or 8 million, different chromosome assortments.**

    b. **Crossing over** also provides variation by exchanging homologous segments that carry different information (alleles).

2. **Fertilization** is random, meaning that the 8 million possible eggs can combine with any of the 8 million possible spermatozoa to give **70 trillion diploid combinations.**

TABLE 19-4. Comparison of Events in Meiosis I and Mitosis

	Meiosis I	Mitosis
Prophase	Homologs synapse (pair) and cross over	No synapsis; no crossing over
Metaphase	Paired homologs line up on plate	Individual chromosomes line up on plate
Anaphase	Centromeres remain intact; homologs separate	Centromeres divide; sister chromatids separate

V. Mutations

A. Definitions and types. Mutations are changes in the genetic material of an organism. Mutations that affect gene function can result in a new phenotype. These are heritable changes, and they are a continuing source of variability in populations.

1. **Some mutations involve a change in the number of chromosomes.** Such changes result from an abnormal meiotic event called **nondisjunction.** In this process, a homologous pair of chromosomes fails to separate during anaphase I or II. This results in one gamete having both copies of the chromosome, and the other gamete having none.

    a. **Aneuploidy,** which is an abnormal chromosome number, occurs when a gamete created by nondisjunction (a gamete receives two of the same type of chromosome or no copy) gets fertilized. Aneuploidy is transmitted to all cells by mitosis, and usually results in miscarriage of a fetus.

    b. **Polyploidy** occurs when there are too many chromosomes in a gamete. For example, a nondisjunction that produces an extra copy of chromosome 21 (trisomy 21) results in a fetus with Down syndrome.

2. **Most mutations result in changes in DNA sequence.**

    a. Mutations can be large, involving large regions of DNA. Large mutations may alter chromosome structure and, therefore, be detected as an **abnormal karyotype.**

    b. Mutations can be as small as a single nucleotide change in the DNA. Such changes are called **point mutations.**

B. Mechanisms of Chromosome Alterations (Figure 19-5)

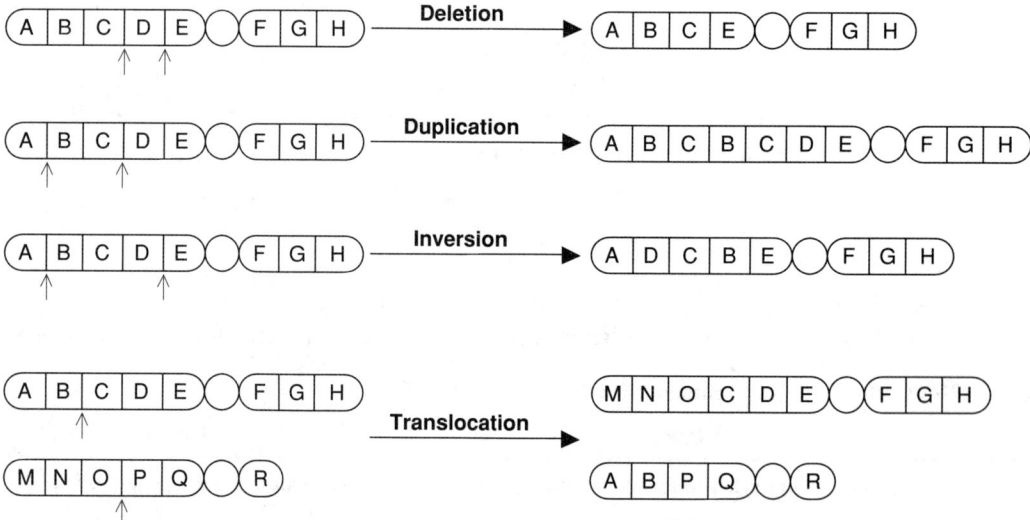

Figure 19-5. Mechanisms of chromosome alterations.

1. **Breaks in a chromosome** can cause a loss or gain of DNA information.
    a. A **deletion** is the loss of DNA sequences. A deletion within a gene almost always negatively affects the phenotype.
    b. A **duplication** is the presence of an extra copy of a sequence on the same chromosome.
2. Chromosome breaks are often accompanied by **rearrangements of DNA sequences.**
    a. An **inversion** is when a sequence reattaches in the reverse orientation.
    b. A **translocation** involves the joining of DNA from one chromosome onto another chromosome.
    c. **Amplification** refers to the production of many extra copies of a sequence. Amplified DNA may be found in the chromosome or on extrachromosomal elements.
    d. When a chromosome is rearranged, the expression of genes near the breakpoints may be influenced by new neighboring sequences. This phenomenon is called **position effect.**

C. **Effects of Mutations on Proteins**
1. A **reading frame shift** is when a mutation changes the codon sequence in the DNA. This results in a new sequence of amino acids past the point of the mutation.
2. A **missense mutation** results in the substitution of one amino acid for another.
3. A **nonsense mutation** creates a new stop codon, resulting in premature termination of a polypeptide chain.
4. **Silent mutations** have no detectable effect on phenotype.
    a. A mutant codon may code for the same amino acid as the normal codon (i.e., redundant codons).
    b. A mutation may occur between genes or in an **intron.**
    c. A mutation may alter a protein with a nonessential or redundant function.

VI. Sex Linkage

A. **Sex Determination**

Sex is a genetically determined trait. In humans, males and females have 22 homologous pairs of chromosomes, or **autosomes.**

1. **Sex chromosomes** are the twenty-third pair of chromosomes.
    a. **Females** have two X chromosomes (**XX**).
    b. **Males** have one X and one Y chromosome (**XY**).

TABLE 19-5. Determination of Maleness in Humans

		Spermatozoa	
		X	**Y**
Eggs	X	XX	XY
	X	XX	XY

The ratio of female to male offspring is 1:1.

2. Even though the **X and Y chromosomes** pair during meiosis I, they **are not homologous.** The Y is small and carries few genes; the X is large and carries many essential genes.

3. In humans, male is the **heterogametic sex;** that is, it is the sex that produces two different kinds of gametes and determines the sex of the offspring (Table 19-5). Maleness in humans is determined by a specific area on the Y chromosome called the **testes-determining factor.** However, mechanisms of sex determination vary in different organisms. In bees, for example, unfertilized eggs (haploid) become males, while fertilized eggs (diploid) become females.

B. Sex-linked Genes

1. Sex-linked genes are genes on the X chromosome. These genes have special **patterns of inheritance** because of the different sex chromosomes in males and females.

    a. Sons receive the X chromosome only from their mothers and the Y chromosome only from their fathers. All daughters, but no sons, receive their father's X chromosome.

    b. **Males only have one X chromosome.** Therefore, any sex-linked recessive gene (received from the mother) is expressed in the male phenotype.

    c. **Females have two X chromosomes** and need two copies of a recessive gene to show the phenotype.

    d. **More males than females have recessive, sex-linked disorders,** such as color blindness, hemophilia, and Duchenne muscular dystrophy.

2. **Males and females both have two copies of the autosomal genes.** However, females have two copies of sex-linked genes and males only have one.

    a. **Cells in female mammals compensate** for this inequality by **inactivating one of the X chromosomes** during embryogenesis, leaving only one active copy of the sex-linked genes.

    b. The inactive X chromosome contracts into a dense body called a **Barr body.**

    c. **Inactivation of X chromosomes is random** in each cell. For heterozygous sex-linked alleles, therefore, each allele is active in about one half of the female's cells.

VII. Pedigree Analysis

A. Pedigrees show a **family's history** for the inheritance of a particular trait. Pedigrees are used to analyze the segregation of inherited disorders in humans.

1. In determining a pedigree, **males are symbolized as squares and females as circles.** Horizontal lines represent mating partners, and vertical lines connect offspring to their parents.

2. **Genotypes** are shown next to the symbols. An individual with the disorder is shown by a solid or crossed square or circle.

B. To determine the **inheritance pattern** in humans, whether the inherited disorder is dominant or recessive, the number of unafflicted individuals is compared with the number of afflicted individuals in the pedigree.

1. **Recessively inherited disorders** are indicated by a 3:1 ratio of unafflicted to afflicted family members (Figure 19-6). Genotypes are determined by assessing family members above and below the afflicted individual in the pedigree; parents and offspring must be heterozygous.

   a. **Recessive disorders often skip generations** because the heterozygotic family members, or **carriers,** are phenotypically normal.

   b. When **two carriers mate,** each offspring has a **1 in 4 chance of being homozygotic** and having the disorder. Each offspring has a 3 in 4 chance of being normal (2 of 3 of these offspring are carriers themselves).

   c. Examples of recessively inherited disorders include phenylketonuria, cystic fibrosis, Tay-Sachs disease, and sickle-cell anemia.

2. **Dominantly inherited disorders** are indicated by a 1:1 ratio in a family (Figure 19-7). A single copy of the dominant gene is sufficient to significantly affect phenotype. Dominant traits **do not usually skip generations.**

   a. Family members who are **heterozygotic for lethal dominant traits do not usually reproduce,** so these genes are less common than lethal recessive genes in a population.

   b. The exception is **late-acting dominant lethal genes,** which show up later in life, after an individual has reproduced. Each offspring has a 1 in 2 chance of receiving the dominant gene. Examples include Huntington chorea and Alzheimer disease.

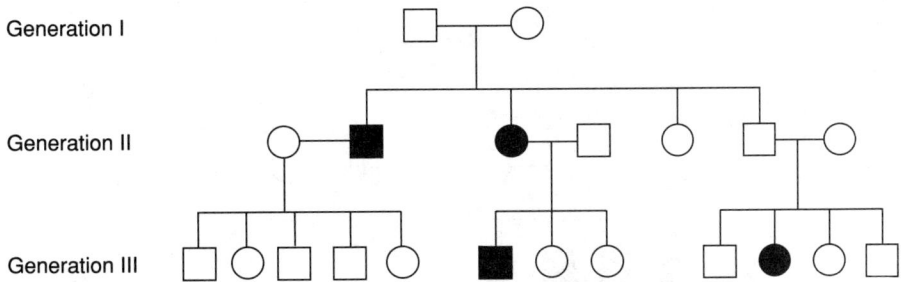

Figure 19-6. A pedigree demonstrating a recessively inherited disorder.

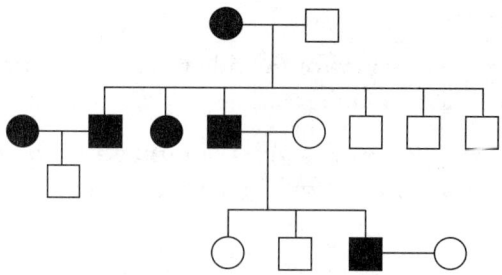

Figure 19-7. A pedigree demonstrating a dominantly inherited disorder.

3. **A sex-linked recessive trait affects more males than females** (Figure 19-8). Such traits are passed from a normal carrier female to all of her sons, who are afflicted. The trait is then passed from the afflicted sons to all of their daughters, who are carriers.

VIII. Gene Mapping

A. Genetic Linkage

1. **Linked genes** are genes that are located on the same chromosome. Linked genes are an **exception to Mendel's law of independent assortment.**

   a. Genes that are adjacent to each other on the chromosome are inherited as a unit (i.e., they go through meiosis and fertilization together). Such genes do not assort independently.

   b. Most of the offspring have the parental genotype at closely linked loci.

2. Genes on the same chromosome but some distance apart can be separated by **crossing over** during meiosis. A crossover can result in exchange of information between paired homologous chromosomes (Figure 19-9).

   a. If the homologs carry different alleles of the exchanged genes, the result is a **new combination of alleles** on the chromosome. This recombination event results in a **recombinant chromosome.**

   b. The **frequency of recombination** between two genes reflects the frequency of crossing over, which is proportional to the distance between the genes.

   c. Recombination events result in production of recombinant offspring.

B. Mapping Genes

1. **Gene mapping** can be performed when recombination frequencies are used to assign a gene to a particular chromosome and to a region of the chromosome. Recom-

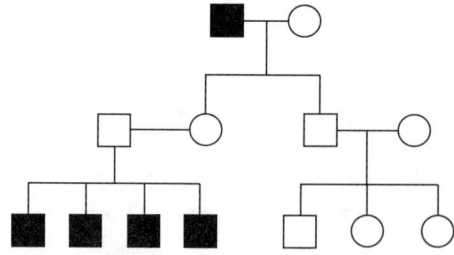

Figure 19-8. A pedigree demonstrating the inheritance of a sex-linked recessive disorder.

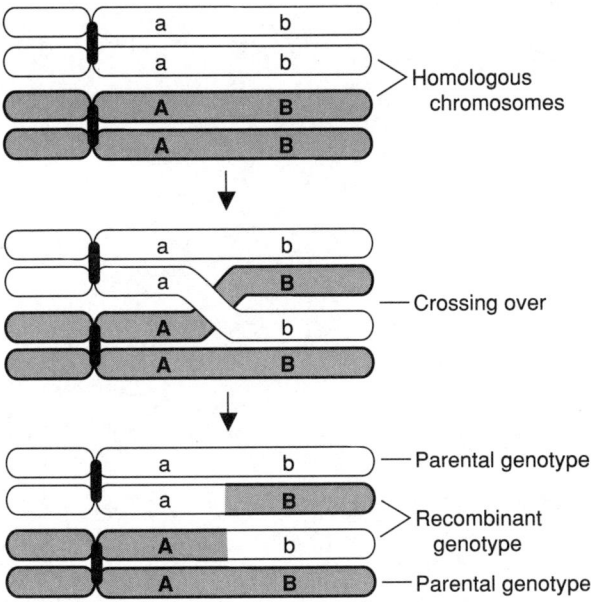

Figure 19-9. The process of crossing over.

bination frequencies can also be used to determine the distance between linked genes.

2. The following is an example of mapping the genes for body color and wing shape in *Drosophila*.

   a. The **dihybrid testcross** is shown in Figure 19-10.

      (1) A recombination frequency of 1% is defined as **1 map unit** on a chromosome.

      (2) Therefore, b and vg are 22 map units apart.

   b. The greater the distance between two genes on the chromosome, the greater the chances of a crossover between them, and the greater the recombination frequency.

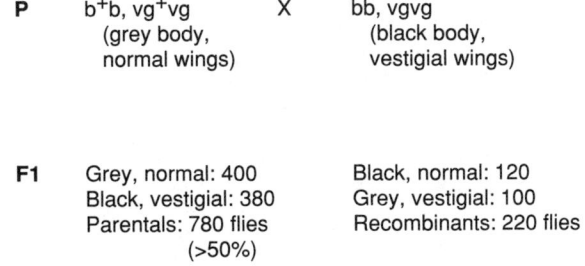

Figure 19-10. A dihybrid testcross used to map the genes for body color and wing shape in *Drosophila*. The recombination frequency of 1% is 1 map unit. Therefore, this testcross demonstrates that the genes b and vg are 22 map units apart.

Evolution   20

I. Natural Selection

A. Charles Darwin

In the 1850s, Charles Darwin published *On the Origin of Species by Means of Natural Selection*. In this book, Darwin presented evidence that species present today evolved from ancestral species. Darwin also proposed a mechanism for evolution that he called **natural selection.**

1. The **theory of natural selection** is based on the following three major concepts:

    a. In any population, **more individuals are produced each generation than can be supported by the environment.** This leads to competition for the limited natural resources and survival of only a fraction of individuals each generation.

    b. The likelihood of survival is not random but depends in part on the fitness of an individual, or **survival of the fittest. Fitness** refers to the inherited traits that make an individual suited to the surrounding environment.

    c. **Individuals who are better suited to their environment are likely to leave more offspring than less fit individuals.** This **differential reproduction** among individuals of a population leads to the gradual accumulation of favorable traits in the population. The gene pool changes over time as the population adapts to its environment.

2. There are **three major modes of natural selection.**

    a. **Stabilizing selection** involves selection of more intermediate rather than extreme phenotypes. This type of selection reduces variability in a population and is seen in relatively stable environments.

    b. **Directional selection** involves selection of a particular phenotypic trait during a time of environmental change.

    c. **Diversifying selection** involves selection of the extreme phenotypes in a population.

B. Nonadaptive Mechanisms of Population Change

1. **Genetic drift** is a chance change in the gene pool of a population. Genetic drift is likely to occur only in very small populations.

a. The **bottleneck effect** is one type of genetic drift in which a random event, such as a disaster, drastically reduces the size of a population.

b. The **founder effect** occurs when a small group of individuals from a parent population colonizes a new area and starts a new population.

2. **Gene flow** involves the migration of individuals between populations.

3. **Nonrandom mating** (inbreeding) is usually based on proximity.

II. The Concept of Species

A. Definition

A species is a **group of individuals with the potential to interbreed** in nature. This mating produces viable, fertile offspring.

B. Reproductive Isolation

Individuals of different species are said to be reproductively isolated. The different mechanisms of isolation include the following:

1. **Geographic barriers** between species

2. **Temporal isolation** because species mate at different times

3. **Physiologic isolation** because of mechanical incompatibility between individuals of different species

4. **Ecological and environmental isolation** because species inhabit different niches

C. Speciation. Speciation is the formation of one or more new species from a previously existing species.

1. There are **two main modes of speciation.**

   a. In **phyletic speciation (anagenesis),** an ancestral species evolves over time into a new species.

   b. In **divergent speciation (cladogenesis),** an ancestral species splits, forming more than one new species over time.

2. There are **three major mechanisms of speciation** based on biogeographical criteria.

   a. In **allopatric speciation,** a geographic barrier isolates two populations, which then evolve independently.

   b. **Sympatric speciation** occurs because a genetic change in a subgroup of a population causes it to become reproductively isolated from the rest of the population. Thus, the gene pool is divided without geographic separation.

   c. **Parapatric speciation** involves the gradual divergence among members of a population to form different species. Parapatric speciation also does not require geographic separation.

III. The Origin of Life

A. Chemical Evolution

Earth is approximately 4 billion years old. The first living organisms are proposed to have arisen by a gradual process of chemical evolution. This process is thought to have involved four stages.

1. **Abiotic (nonliving) synthesis.** First, organic compounds were synthesized from inorganic precursors available in the atmosphere and seas of the primitive Earth (e.g., water, methane, ammonia, hydrogen gas). This process of abiotic (nonliving) synthesis led to the accumulation of small organic compounds.

2. **Formation of organic polymers.** Organic compounds were joined together to form organic polymers.

3. **Formation of protobionts.** Organic molecules aggregated together to form droplets called protobionts. It is thought that protobionts had some of the properties of living cells, such as metabolism, excitability, and a distinct internal environment.

4. **Evolution.** The evolution of heredity began with the origin of genetic information.

B. Earliest Fossil

According to the fossil record, the earliest organisms were prokaryotic cells that arose about 3 billion years ago.

C. Classification

All organisms are grouped into five **kingdoms** (Figure 20-1). Organisms within the same taxonomic categories are related by their evolutionary history (phylogeny).

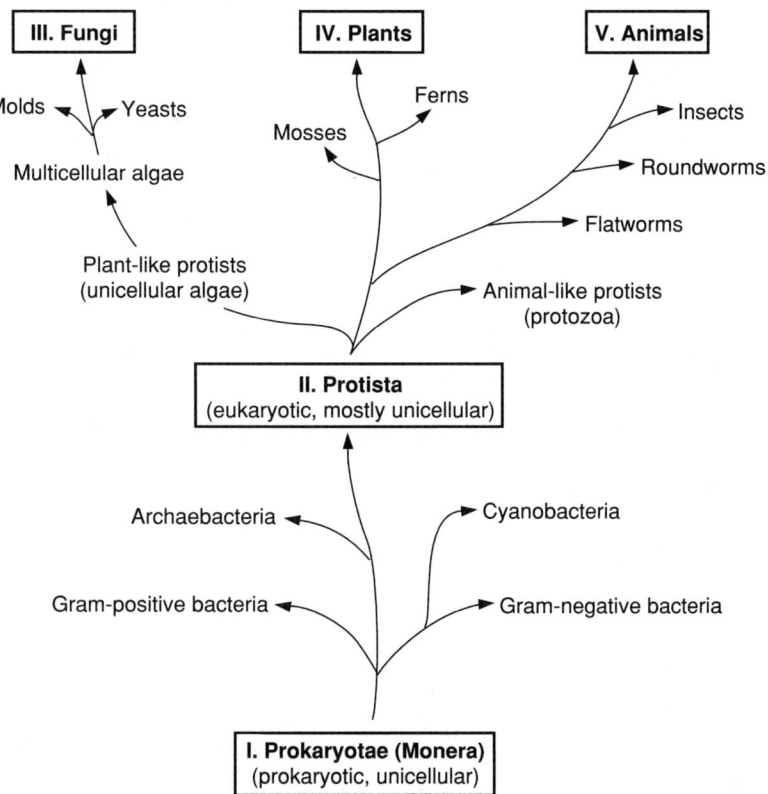

Figure 20-1. The five kingdoms used in taxonomic categorization.

IV. Comparative Anatomy

Anatomic similarities between organisms in the same taxonomic group provide evidence for evolution from a common ancestor.

A. **Homology** is similarity due to common ancestry. An example of homologous structures are the forelegs, wings, flippers, and arms of chordates. Although evolved for different functions, all of these structures are built from similar skeletal elements.

1. Closely related organisms that look different in adults go through **similar stages in their embryonic development.** For example, all vertebrates go through similar embryonic stages, including a stage in which they have gill slits. These develop into gills in fish and into various other structures in different vertebrates.

2. The idea that embryonic development is a replay of an organism's evolutionary history is referred to as **"ontogeny recapitulates phylogeny."**

B. **Analogy** is similarity between species that are not evolutionarily related. Analogous structures may evolve independently because of some common selective advantage. This type of evolution is called **convergent evolution.** For example, the wings of insects and birds have evolved independently. Although they are both used for flying, they are built from entirely different structures.

An Overview of Botany (21)

I. Plant Evolution

A. Plants most likely evolved from water-living green algae **(protists)**. Multiple structural similarities are found between the green algae and plants. Both:

1. Have cell walls made of cellulose

2. Contain chlorophyll *a* and chlorophyll *b*

3. Store energy in the form of starch

4. Demonstrate a similar pattern of thylakoid disk orientation in chloroplasts (i.e., grana)

B. There are **four important periods of plant evolution.**

1. **Adaptation to land.** The evolution of land-dwelling plants probably occurred from green algae approximately 425 million years ago. Mosses are an example of these early plants. Early evolutionary adaptations to land included:

    a. A cuticle to prevent drying in air

    b. Vascular tissue (in some species)

2. **Diversification of vascular plants** (e.g., ferns) occurred approximately 400 million years ago.

3. **Evolution of seeds** occurred approximately 360 million years ago. Seeds provided a significant advance for several reasons. They:

    a. Protected the plant embryo from dessication

    b. Contained energy stores for the embryo

    c. Provided enhanced survival for plant embryos

4. **Evolution of flowering plants.** The fourth period of plant evolution gave rise to the flowering plants (angiosperms) approximately 125 million years ago. **Pollination** evolved as a new way for sperm to reach eggs, enabling sexual reproduction to occur in non-moist environments. The angiosperms comprise the majority of the plant species living today.

II. Plant Anatomy

A. Cellular structure. Like animal cells, plant cells contain a nucleus, ribosomes, Golgi bodies, mitochondria, endoplasmic reticulum, microtubules, and microfilaments. In addition, they possess several specialized organelles and structures:

1. **Chloroplasts,** the organelles in which photosynthesis occurs.

2. A **central vacuole** which is involved in nutrient storage and plant growth.

3. A **cellulose-containing cell wall** that provides protection and support to the plant cell. In some plant cells, the wall is a single layer (i.e., a primary cell wall), whereas in others, it is a double layer (i.e., a primary and secondary cell wall).

4. **Plasmodesmata** are channels that connect the cytoplasm of adjacent plant cells.

B. Roots (Table 21-1; Figure 21-1). Water and minerals enter the root through the epidermis, cross the cortex, enter the stele region, and flow into the xylem, which transports the water and minerals into the upper regions of the plant. Transport is mediated by **cohesion-adhesion forces** and **transpiration.**

1. **Cohesion-adhesion forces**

   a. **Cohesion.** Water molecules are attracted to one another by **cohesion,** which is mediated by hydrogen bonding. Cohesion prevents water molecules from separating.

   b. **Adhesion.** Water molecules are strongly attracted to the hydrophilic walls of the xylem cells, giving rise to **adhesive forces.** Surface tension also acts on the water column, leading to meniscus formation.

2. **Transpiration** involves the loss of water through the microscopic pores of leaves. The net effect of transpiration is to exert an "upward pull" on the water column in the xylem.

TABLE 21-1. Structure and Function of the Components of Plant Roots

Root Part	Description	Function
Epidermis	Outermost layer of root	Protection
Root hairs	Extend from epidermis	Increase surface area for absorption
Cortex	Widest area of root, located between the epidermis and endodermis	Stores food
Endodermis	Single layer of cells between the cortex and stele	Separates cortex and stele
Stele	The central vascular region of the root	Contains the xylem and phloem
Xylem	Thick-walled cells in stele	Transports water up the root
Phloem	Thin-walled cells in stele	Transports food up the root
Apical meristem	Group of cells located just above the tip of a growing root	Supplies cells for root to grow in length
Root cap	Covers tip of root	Protects the meristem as the root elongates through the soil

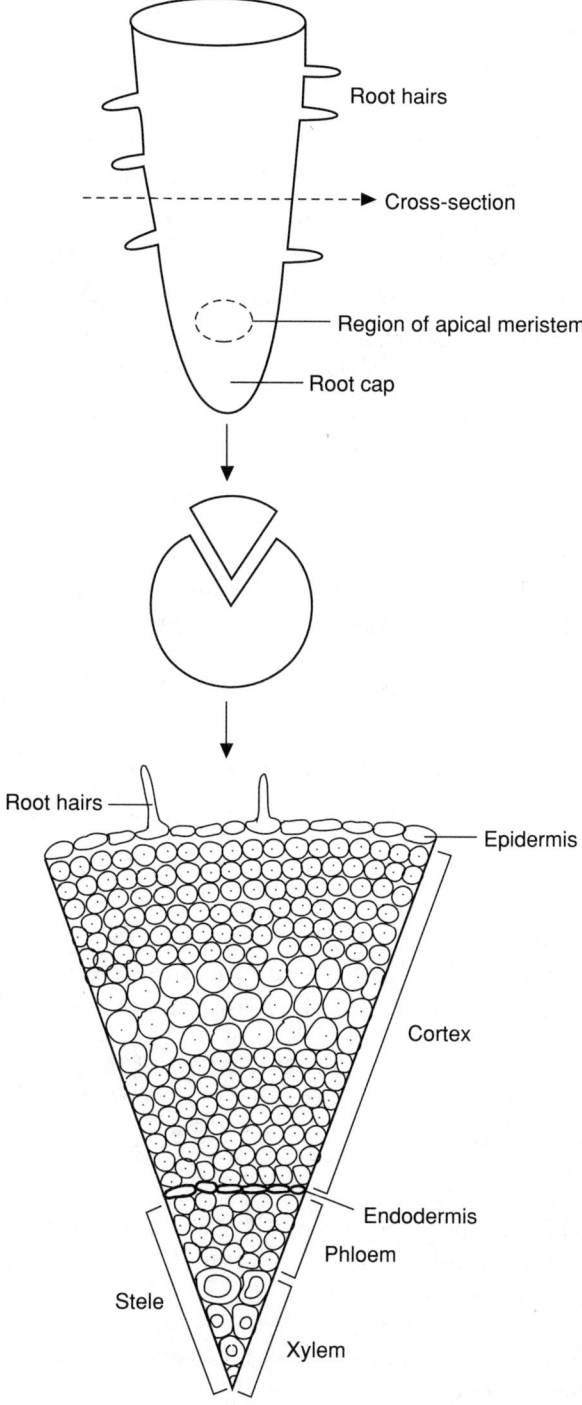

Figure 21-1. Cross-sectional view of a simple plant root.

TABLE 21-2. Structure and Function of the Components of Plant Stems

Stem Part	Description	Function
Cork/epidermis	Outermost layer	Protects against water loss and trauma
Cork cambium*	Single cell layer just inside the cork layer of the tree stem	Produces new cork cells
Cortex	First layer inside the cork cambium	Stores food
Pith	Central tissue	Stores food
Xylem	Located external to the pith layer	Transports water and provides stem support
Phloem	Located inside the cortex layer	Transports food
Vascular cambium	Single cell layer at top edge of xylem	Produces new phloem and xylem cells
Bark*	The outermost layers of a tree trunk, comprised of living phloem, cork cambium, and cork layers	Protection

*Present only in tree stems (trunks).

C. Stems

1. **Anatomy** (Table 21-2; Figure 21-2). The anatomy of a stem varies to some extent, depending on the type of plant (e.g., trunks versus flower stems).

    a. Plant stems have **nodes** where leaves are attached. **Internodes** are the segments of stem between nodes.

    b. **Buds** form leaves. **Axillary buds** grow from the side of the stem, whereas **terminal buds** grow at the end of a stem.

2. **Growth**

    a. **Lengthening.** Meristems are centers of growing cells found in the stem, buds, and roots of plants. **Shoot apical meristems** allow plants to grow taller through a process known as bud growth.

    b. **Widening.** The stems or the trunk of plants grow in width by **secondary growth.** In this process, the lateral meristem (the vascular cambium in tree trunks) produces secondary xylem and phloem, and the cork cambium produces additional cork.

    (1) The secondary xylem is produced yearly and forms wood. The rings one can see in split wood show the yearly activity of secondary xylem formation.

    (2) The secondary phloem does not accumulate yearly. It develops into bark and is sloughed.

D. Leaves

1. **Size and shape.** A leaf is **simple** if it has a single, undivided surface (blade). If the surface of a leaf is divided into several leaflets, it is termed **compound**.

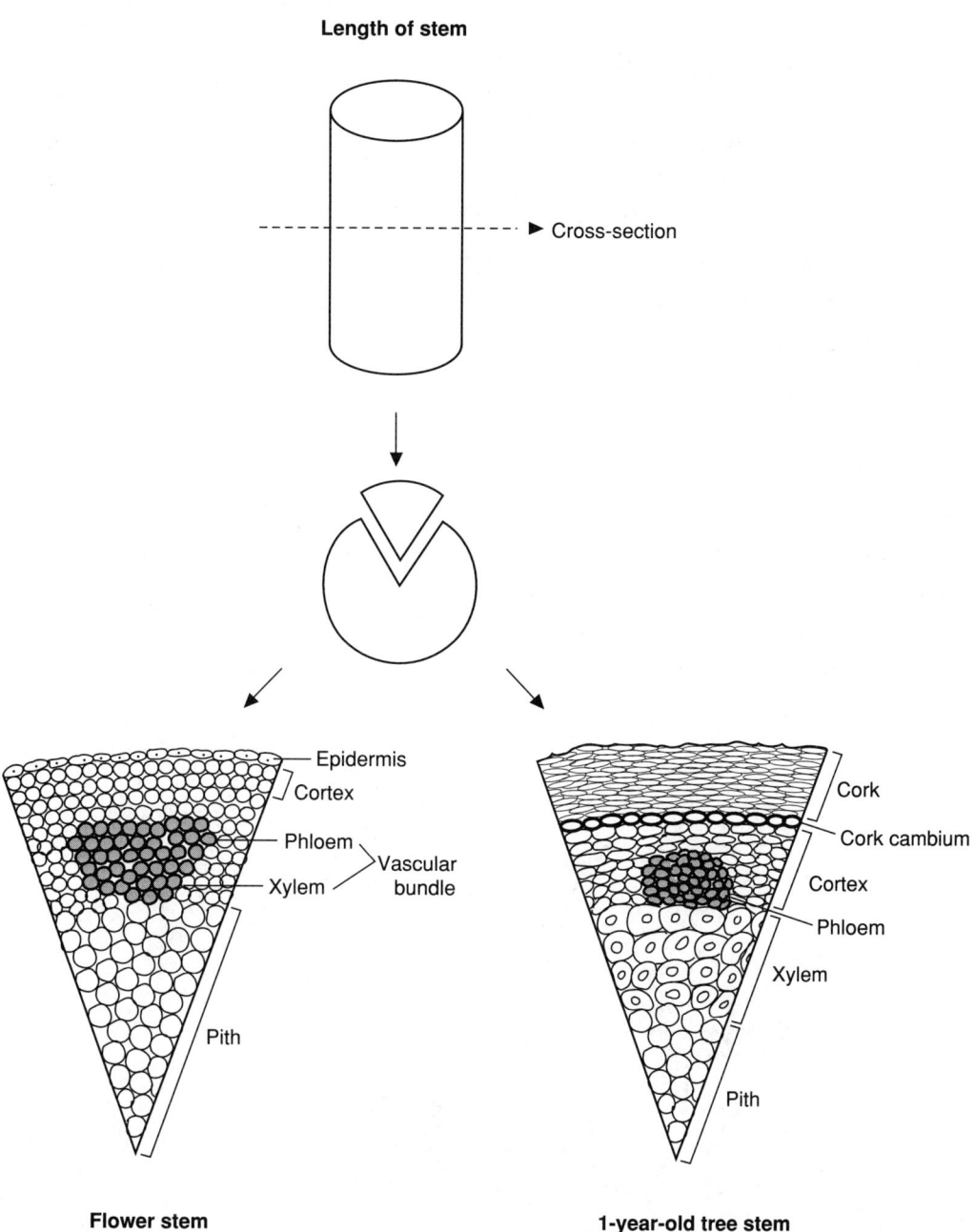

Figure 21-2. Cross-sectional views of a flower stem (*left*) and a 1-year-old tree stem (*right*). (Not drawn to scale.)

2. **Basic tissue types**
   a. The **epidermis** is a waxy cuticle that protects against water loss and physical damage. It is one cell-layer thick and covers the outer surface of both sides of a leaf. Tiny pores (**stomata**) penetrate the epidermis to allow gas exchange between the air and the photosynthetic cells in the leaf.
   b. The **vascular tissue** contains a complex network of **veins.** These veins deliver water and minerals from the xylem of the stem to the photosynthetic cells of the mesophyll. The veins also deliver sugars and other photosynthetic products to the phloem for transport to other regions of the plant.
   c. The **mesophyll** contains photosynthetic cells and is in close contact with the vascular tissue.

E. **Flowers.** (Figure 21-3). There are four important anatomical regions of a flower:
   1. **Septals** protect the floral bud before it opens. For example, septals are the outer green structures covering closed rose buds.
   2. **Petals,** the colored parts of a flower, attract pollinators (e.g., bees, insects).
   3. The **stamen** produces **pollen grains** (male gametophytes). The stalk of the stamen is known as the **filament,** and the head of the stamen is known as the **anther.**
   4. The **carpel** has a sticky upper portion known as the **stigma,** a slender neck known as the **style,** and a base known as the **ovary.** Within the ovary are one or more **ovules,** each of which contains an **egg cell** (female gametophyte).

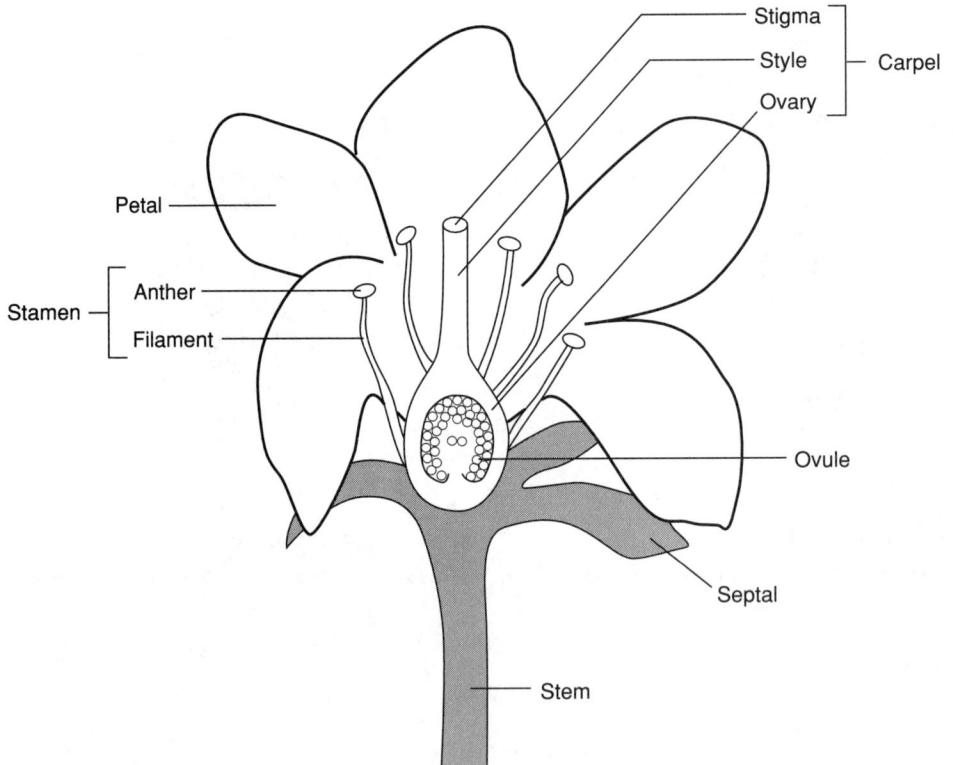

Figure 21-3. The structure of a flower. The internal structure of the ovary is shown.

III. Life Cycle

Many plants exist in two different forms during their life cycle. When plants alternate between these forms, **alternation of generations** is said to occur.

A. The **gametophyte** form of a plant produces gametes. In simple plants, the gametophyte form is generally moderate in size, while in complex plants, the gametophyte form is small.

B. The **sporophyte** form of a plant produces spores. In simple plants, the sporophyte form is generally moderate in size, while in complex plants, the sporophyte form is larger and is predominant.

IV. Kingdom Plantae

Plants occupy one of the five major kingdoms of living things, the **kingdom plantae.** There are twelve divisions within the plant kingdom (Table 21-3).

A. Nonvascular plants

    1. General characteristics

        a. All of these plants have a **waxy cuticle.**

        b. The gametes develop in a protective organ, the **gametangium.**

            (1) The **male gametangium** produces **sperm,** which have flagella, are motile, and require a moist environment to swim toward the egg.

            (2) The **female gametangium,** the **archegonium,** produces the **egg.** Fertilization occurs in the archegonium and leads to the development of an embryo.

TABLE 21-3. Divisions of Kingdom Plantae

Division Name	Common Name
Nonvascular plants	
Division Bryophyta	Mosses
Division Hepatophyta	Liverworts
Division Anthocerophyta*	Hornworts
Vascular plants, seedless	
Division Psilophyta*	Wiskferns
Division Lycophyta	Club mosses
Division Sphenophyta*	Horsetails
Division Pterophyta	Ferns
Vascular plants, with seeds	
Gymnosperms	
Division Coniferophyta	Conifers
Division Cycadophyta*	Cycads
Division Ginkgophyta*	Ginkgo
Division Gnetophyta*	Gnetae
Angiosperms	
Division Anthophyta	Flowering plants

* Contains fewer than 100 species

c. Because the nonvascular plants do not contain a vascular system, they use **diffusion, capillary action,** and **cytoplasmic streaming** to distribute water throughout the plant.

2. **Division Bryophyta** includes the **mosses.** There are over 10,000 species.

   a. Mosses have root-like structures (**rhizoids**) and grow in mat-like groups.

   b. The life cycle alternates haploid and diploid generations.

      (1) The **haploid plant** is known as a **gametophyte.** The haploid gametophyte is the dominant generation.

      (2) The **diploid plant** is known as a **sporophyte.** The diploid sporophyte produces haploid spores via meiosis. The spores develop into new gametophytes.

3. **Division Hepatophyta** includes the **liverworts.** There are over 6000 species.

   a. These plants are small and divided into lobes, hence the name (i.e., after the liver, a lobated organ).

   b. The reproductive cycle is similar to that of the mosses. Liverworts may also reproduce asexually.

4. **Division Anthocerophyta** includes the **hornworts.** There are fewer than 100 species. These plants are small like liverworts, but have long sporophytes that look like horns.

B. Vascular plants (seedless)

1. **Division Psilophyta** includes simple, ancient plants known as **whiskferns.** These plants have many stems but no true leaves.

2. **Division Lycophyta** includes **club mosses (ground pines).** These are common names; however, Lycophytes are neither mosses nor pines. Rather, they are small plants that live on forest floors or grow in the tropics. Many tropical Lycophytes are **epiphytes** (i.e., plants that grow on another plant without damaging it); they grow on trees.

3. **Division Sphenophyta** includes **horsetails,** which are also ancient plants. Only about 20 species still survive. Horsetails have a straight, jointed stem and a cone-like tip and live in damp areas.

4. **Division Pterophyta** includes the **ferns.** These are the most successful modern-day seedless vascular plants. Over 10,000 species exist.

   a. Ferns are found in the tropics and in temperate regions. Ferns generally require moist environments for survival.

   b. Ferns have haploid and diploid life cycles.

      (1) Haploid fern spores (from mature fern plants) develop into small gametophyte plants.

- (2) Gametophytes produce eggs and sperm.

- (3) Fertilized eggs develop into sporophytes, which develop into fern plants (mature sporophytes).

- (4) The fern plant develops clusters of spores under its leaves, which ultimately give rise to new gametophytes.

C. Vascular plants (with seeds)

1. **Gymnosperms**

   a. **Division Coniferopyta** includes the conifers (e.g., redwoods, pines, cedars, firs, juniper), evergreens with cones and needle-shaped leaves. The typical life cycle of a conifer includes haploid and diploid stages. Two types of cones develop, **pollen cones** and **ovulate cones.**

      (1) Haploid spores, produced by meiosis, are released from pollen cones, leading to the pollination of ovules (contained in ovulate cones).

      (2) Pollen grains enter individual scales of ovulate cones and slowly grow a pollen tube that enlarges toward the egg cell. Haploid sperm move through the pollen tube to the egg and fertilize it. A new embryo (sporophyte) then develops into a conifer seedling plant.

   b. **Division Cycadophyta** (palm-like plants), **Division Ginkgophyta** (maidenhair trees), and **Division Gnetophyta** (an unusual group of diverse plants with tree, vine, or shrub characteristics) are also gymnosperms.

2. **Angiosperms** are the most diverse and numerous plants. Nearly 250,000 species are known.

   a. Angiosperms have flowers and fruit. Fruit, a mature ovary that contains seeds, evolved as a mechanism for seed dispersal. Animals consume fruit and digest the edible part, while the seeds pass through the digestive tract undamaged. Seeds are then deposited in feces some distance from the parent plant.

   b. Angiosperms demonstrate a complex **alternation of generations life cycle**, with **haploid, diploid,** and **triploid stages.**

      (1) Haploid pollen develops in the stamen of the flower and is carried by wind or pollinators (insects) to the carpel.

      (2) Pollen adheres to the stigma and sends a pollen tube down the style toward the ovary.

      (3) The pollen tube ultimately penetrates the ovule and releases **two** haploid sperm. One sperm fertilizes the egg, producing a diploid zygote. The other sperm fuses with two nuclei in the embryo sac and forms a triploid nucleus. This process is known as **double fertilization.**

(4) The triploid nucleus forms **endosperm,** a tissue rich in food reserves (starch). The zygote forms a sporophyte embryo. The entire ovule forms a seed (ovule).

(5) The ovule contains the embryo, endosperm, and an outer protective seed coat. The seed ultimately germinates and produces a sporophyte seedling.

HIGH-YIELD REVIEW QUESTIONS

Section I: Molecular and Cellular Biology

Question Set 1

1. L-aromatic amino acid decarboxylase (AAAD) catalyzes the conversion of tyrosine to dopamine:

   Tyrosine → (AAAD) → Dopamine

   If 1 mole of tyrosine is added to 10^{-10} moles of AAAD, dopamine production proceeds at a rate of 10^{-8} moles/min. If 10^{-3} moles of phenylalanine are then added, dopamine production falls to 10^{-9} moles/min. If more tyrosine is added, dopamine production increases. The most likely effect of phenylalanine is:

   A. allosteric inhibition of AAAD.
   B. irreversible inhibition of AAAD.
   C. competitive inhibition of AAAD.
   D. noncompetitive inhibition of AAAD.

2. The enzyme COMT catalyzes the conversion of epinephrine to its inactive metabolite, methylepinephrine. If 10^{-8} moles of epinephrine are added to 10^{-14} moles of COMT, methylepinephrine production proceeds at a rate of 10^{-12} moles/sec. If 10^{-12} moles of vitamin C are then added, the rate of metabolite production increases to 10^{-11} moles/sec. Vitamin C is probably acting as a(n):

   A. allosteric activator of COMT.
   B. competitive inhibitor of COMT.
   C. noncompetitive inhibitor of COMT.
   D. enzyme denaturing agent.

3. Lactic acid is generated as an end product of anaerobic metabolism. The gland or organ that removes lactate from the circulation and resynthesizes it to acetyl CoA is the:

   A. liver.
   B. spleen.
   C. kidney.
   D. pituitary.

4. What is the ultimate fate of the glucose carbon chain in metabolism?

   A. ATP
   B. CO_2
   C. O_2
   D. NADH

5. The production of ATP from the energy stored in NADH depends on:

   A. hydrolysis.
   B. oxidation/reduction.
   C. chemotaxis.
   D. dephosphorylation.

6. Which process does NOT occur in intracellular organelles?

   A. Oxidative phosphorylation
   B. Transcription
   C. Electron transport
   D. Glycolysis

7. In the absence of molecular oxygen, which process does NOT continue?

   A. Glycolysis
   B. Conversion of pyruvate to lactate
   C. ATP production
   D. Electron transport

8. Skin infections are most commonly caused by spherically shaped bacteria found in clusters. The most likely name for these bacteria is:

   A. rods.
   B. spirilla.
   C. streptococci.
   D. staphylococci.

Questions 9–11 refer to a sequence of DNA found on one strand of a double-stranded DNA molecule. This DNA sequence is:

3´AAGGCTAGC5´

9. What is the complementary DNA strand sequence?

   A. 5´AAGGCTAGC3´
   B. 3´TTCCGATCG5´
   C. 5´TTCCGATCG3´
   D. 5´UUCCGAUCG3´

10. What is the primary mRNA transcript made from the complementary strand identified in question 9?

    A. 5´CGATCGGAA3´
    B. 5´CGAUCGGAA3´
    C. 5´UUCCGAUCG3´
    D. 5´GCUAGCCUU3´

11. What is the tRNA anticodon sequence coded for by the original DNA sequence?

    A. AAG GCT AGC
    B. TTC CGA TCG
    C. AAG GCU AGC
    D. UUC CGA UCG

12. Which statement about both eukaryotes and prokaryotes is false?

    A. Both have double-stranded DNA.
    B. Both have single-stranded RNA.
    C. Both use free ribosomes as sites of protein synthesis.
    D. Both replicate their chromosomal DNA from one unique origin.

13. Which general statement(s) about genetic material is (are) true?

    I. It can be double-stranded DNA.
    II. It can be single-stranded RNA.
    III. It can be single-stranded DNA.
    IV. It can be double-stranded RNA.

    A. I and II
    B. I, II, and III
    C. II and IV
    D. I, II, III, and IV

14. Which structure is NOT important in the synthesis of a novel protein to be inserted into the cell plasma membrane?

    A. Golgi apparatus
    B. Free ribosomes
    C. tRNA
    D. Endoplasmic reticulum

15. Which process is NOT shared by eukaryotes and prokaryotes?

    A. DNA synthesis precedes spindle formation during division.
    B. Each daughter cell receives half of the maternal DNA.
    C. Glycolysis provides ATP for cell use.
    D. Oxidative phosphorylation may provide ATP for cell use.

16. Which is (are) NOT considered fungus (fungi)?

    A. Yeast
    B. Mold
    C. Both yeast and mold
    D. Neither yeast nor mold

17. What is the order of the phases of mitosis?

    A. Interphase, prophase, metaphase, anaphase, telophase
    B. Interphase, telophase, metaphase, prophase, anaphase
    C. Interphase, prophase, anaphase, telophase, metaphase
    D. Prophase, metaphase, anaphase, interphase, telophase

18. Lipid bilayers are NOT found as part of:

    A. mitochondria.
    B. endoplasmic reticulum.
    C. lysosomes.
    D. nucleoli

19. Which statement does NOT describe characteristics of fungi?

    A. They are eukaryotes.
    B. They may be either haploid or diploid.
    C. They may reproduce sexually or asexually.
    D. Because they have no mitochondria, they are capable only of glycolysis.

20. Which combination of biochemical molecules are the primary components of bacterial cell walls?

    A. Protein and carbohydrate
    B. Carbohydrate and lipid
    C. Lipid and protein
    D. Protein and nucleic acid

SOLUTIONS

Molecular and Cellular Biology Set 1

1. **C** Phenylalanine, which is similar in structure to dopamine, acts as a competitive inhibitor of AAAD:

   $\text{C}_6\text{H}_5\text{-CH}_2\text{-CH(NH}_2\text{)-COOH}$

   Phenylalanine

   Recall that competitive inhibitors resemble the normal substrate and compete with it for binding to the active site of an enzyme. Inhibition of AAAD with phenylalanine can be overcome by adding more tyrosine.

2. **A** Vitamin C increases the rate of methylepinephrine production. It most likely acts at a site on the enzyme other than the active site and increases enzyme activity. This action is known as allosteric activation, or positive allosteric control. Choice B is incorrect because competitive processes act at the active site and inhibit rather than augment the production of product. Choices C and D are incorrect because both would decrease product production.

3. **A** The liver removes lactate from the circulation and metabolizes it to acetyl CoA, which is returned to the circulation, in a process known as the Cori cycle. The spleen is mainly involved in storage of old red blood cells. The kidney produces concentrated urine and reabsorbs and excretes solutes. The pituitary produces and releases hormones.

4. **B** Glucose is metabolized to pyruvate in glycolysis. Pyruvate loses one CO_2 to form acetyl CoA. Each acetyl CoA enters the Krebs cycle where it is decarboxylated. Therefore, the ultimate fate of glucose carbon units is CO_2.

5. **B** NADH transfers its electrons to electron transport proteins in the process of oxidative phosphorylation. Recall that oxidative phosphorylation involves the production of ATP using energy from the redox reactions of the electron transport chain. The electron transport proteins have increasing electron affinities, and the phosphorylation of ADP to ATP is associated with electron transfers between proteins.

6. **D** Glycolysis occurs in the cytoplasm of the cell. Oxidative phosphorylation, the Krebs cycle, and electron transport occur in the mitochondrion. Transcription occurs in the nucleus.

7. **D** If the final step of electron transport ceases (the transfer of electrons to molecular oxygen forming H_2O), electron transport stops. Glycolysis, conversion of pyruvate to lactate, and some reactions producing ATP are independent of the presence of molecular oxygen.

8. **D** Staphylococci are spherical bacteria found in clusters. Streptococci are spherical bacteria found in chains or pairs. Rods are rod-shaped bacteria, whereas spirilla are spiral-shaped bacteria.

9. **C**   10. **B**   11. **C**

    The key to understanding these related questions is knowing the base pairing rules. For DNA, G bonds to C, whereas A bonds to T. For RNA, G bonds to C, and A bonds to U. Also note that mRNA is complementary to DNA, and tRNA is complementary to mRNA. Therefore, for this set of questions:

Original DNA	3´AAGGCTAGC5´
Complementary DNA	5´TTCCGATCG3´
mRNA	3´AAGGCUAGC5´

Original DNA	3´AAGGCTAGC5´
mRNA	5´UUCCGAUCG3´
tRNA	AAG GCU AGC

12. **D** Eukaryotes may replicate chromosomal DNA from several origins simultaneously. Choices A–C are true statements.

13. **D** Although eukaryotes and bacteria have double-stranded DNA, viruses have

14. **B** other forms of genetic material. Viruses may have either DNA or RNA as their genetic material. Furthermore, viral DNA or RNA may be either single-stranded or double-stranded, and either linear or circular.

14. **B** The rough endoplasmic reticulum (containing ribosomes) and the Golgi apparatus are generally required for the production of plasma membrane proteins. Recall that these proteins are often glycoproteins, and are glycosylated in the Golgi apparatus. Transcription of mRNA occurs in the nucleus, and translation of proteins requires tRNA. Free ribosomes are not required to synthesize cell membrane proteins because the ribosomes involved in the translation of these proteins are embedded in the endoplasmic reticulum.

15. **A** There is no spindle formation in prokaryotes (bacteria) because bacteria do not use the process of mitosis to reproduce. Prokaryotes reproduce by binary fission.

16. **D** Both yeast and mold are fungi. The three major types of fungi are distinguished by their structure. Molds are multicellular filamentous organisms. Fleshy fungi are multicellular filamentous organisms that produce a thick (fleshy) reproductive body. These fungi include mushrooms. Yeast are nonfilamentous, unicellular organisms.

17. **A** Mitosis is a process of cell division in which diploid (2n) cells equally allocate replicated chromosomes to each of the diploid (2n) daughter cells. Growth occurs during interphase. Active cell division occurs in four additional stages, in the following order: prophase, metaphase, anaphase, telophase.

18. **D** Mitochondria, endoplasmic reticulum, lysosomes, and the nucleus each contain a lipid bilayer. The nucleolus is made of rRNA and protein and does not contain a lipid bilayer.

19. **D** Fungi are eukaryotes and have membrane-bound organelles, including mitochondria. They also may be haploid or diploid and reproduce either sexually or asexually.

20. **A** Bacterial cell walls consist of protein and carbohydrate for the most part. Peptidoglycan, the major framework unit of the cell wall, consists of carbohydrate moieties bonded to small peptide chains.

HIGH-YIELD REVIEW QUESTIONS

Section I: Molecular and Cellular Biology

Question Set 2

1. The active site of an enzyme is determined to be hydrophobic and positively charged. A substrate that readily binds to the active site is more likely:

   A. hydrophilic and positively charged.
   B. hydrophilic and negatively charged.
   C. hydrophobic and positively charged.
   D. hydrophobic and negatively charged.

2. Enzyme cofactors are BEST described as:

   A. nonprotein substances required for enzyme activity.
   B. small, nonprotein, organic molecules.
   C. metal ions or coenzymes that bind tightly to an enzyme.
   D. catalysts.

3. The substrate concentration in an enzyme test system is slowly increased to high levels. As the substrate concentration increases:

   A. there is a continuous increase in the reaction rate.
   B. the increase in the reaction rate slows down.
   C. there is an exponential increase followed by a plateau phase.
   D. there is no change in reaction rate.

4. Which statement about enzymes is NOT true?

   A. Many enzymes show a distinct pH optimum.
   B. Many enzymes show a linear increase in reaction rate with increasing temperature.
   C. Competitive inhibitors compete with substrate for binding at the enzyme active site.
   D. The binding of noncompetitive inhibitors outside the active site distorts the active site.

5. Which statement about enzymes or enzyme-catalyzed reactions is NOT true?

   A. Enzyme reactions are reversible.
   B. Substrate specificity for enzymes is absolute or relative.
   C. Enzyme–substrate bonds are covalent.
   D. Enzymes lower activation energies of spontaneous reactions.

6. The process of respiration does NOT include:

   A. glycolysis.
   B. generation of acetyl CoA from pyruvate.
   C. the Krebs cycle.
   D. oxidative phosphorylation.

7. Approximately what percent of the total ATP generated from the glycolysis and complete respiration of glucose is produced by oxidative phosphorylation?

   A. Less than 15%
   B. Approximately 25%
   C. Approximately 75%
   D. More than 85%

8. What does the net reaction of aerobic glycolysis of 1 mole of glucose to 2 moles of pyruvate yield?

   A. 4 ATP + 2 NADH + 2 H^+
   B. 2 ATP + NADH + H^+
   C. 4 ATP + 4 NADH + 4 H^+
   D. 2 ATP + 2 NADH + 2 H^+

9. Human red blood cells lack mitochondria. Which process is the most likely mechanism by which these cells metabolize glucose?

   A. Glycolysis
   B. The Krebs cycle
   C. Oxidative phosphorylation
   D. All three choices are correct

10. Bioenergetically, for each mole of acetate that it oxidizes, the Krebs cycle produces:

    A. 12 ATP.
    B. 2 moles NADH, 2 moles $FADH_2$, 2 moles GTP.
    C. 3 moles NADH, 1 mole $FADH_2$, 1 mole GTP.
    D. 2 moles NADH, 3 moles $FADH_2$.

11. Which enzyme catalyzes the rate-limiting step in glycolysis?

    A. Hexokinase
    B. Pyruvate kinase
    C. Enolase
    D. Phosphofructokinase

12. In eukaryotic cells, where does the Krebs cycle occur?

    A. The cytoplasm
    B. The mitochondrial matrix
    C. The inner mitochondrial membrane
    D. The outer mitochondrial membrane

13. In cellular metabolism, electrons are removed from the acetyl group in acetyl CoA and incorporated into reduced coenzymes. Which process is described?

    A. Glycolysis
    B. The Krebs cycle
    C. Fermentation
    D. Oxidative phosphorylation

14. One molecule of acetyl CoA, radioactively labeled at each carbon atom, enters the Krebs cycle. After one turn of the cycle, how many molecules of radioactive CO_2 are produced?

    A. 0
    B. 1
    C. 2
    D. 3

15. Which base is NOT a pyrimidine?

    A. Cytosine
    B. Uracil
    C. Thymine
    D. Adenine

16. Which equation about the base content of DNA is NOT true?

    A. A + G = T + C
    B. G = C
    C. A = T
    D. A + T = G + C

17. Which bond links nucleotides to one another in RNA?
    A. Phosphodiester bonds
    B. Glycosidic bonds
    C. Phosphate bonds
    D. Hydrogen bonds

18. A small fragment of DNA has the following sequence: dp-ACGGTAC-p. If 5′ ends are considered to the left of 3′ ends, which base pair is antiparallel with the given DNA fragment?

    A. Dp-TGCCATC-p
    B. Dp-GTACCGT-p
    C. Dp-ACGGTAC-p
    D. Dp-TGCCTAG-p

19. In mammals, DNA is replicated such that offspring DNA contain:

    A. duplexes of parental strands and duplexes of newly synthesized strands.
    B. circular DNA.
    C. duplexes containing one parental strand and one newly synthesized strand.
    D. duplexes of parental strands only.

20. Which feature is NOT involved in human DNA replication?

    A. Okazaki fragments
    B. The creation of primers
    C. Continuous DNA replication
    D. The need for RNA synthesis

21. Of the four enzymes listed, which enzymes participate in DNA replication?

    I. DNA ligase
    II. DNA polymerase
    III. Primase
    IV. Helicase

    A. I, II and IV
    B. I and III
    C. II and IV
    D. I, II, III, and IV

22. Which enzyme initiates DNA replication?

    A. DNA polymerase
    B. Primase
    C. DNA ligase
    D. Endonuclease

23. RNA and DNA have numerous differences in both structure and function. However, RNA and DNA are similar in that they both classically contain:

    A. the same sugar.
    B. the same purines.
    C. complementary antiparallel strands.
    D. the same pyrimidines.

24. Which DNA would tend to be most resistant to denaturation by temperature?

    A. G–C-rich DNA
    B. A–T-rich DNA
    C. Purine-rich DNA
    D. Pyrimidine-rich DNA

25. Transcription involves:

    A. the formation of proteins.
    B. a 3′ to 5′ synthesis of RNA.
    C. synthesis of RNA from either DNA strand.
    D. the formation of messenger RNA from a DNA template.

SOLUTIONS

Molecular and Cellular Biology Set 2

1. **D** The hydrophobic amino acids in the active site of an enzyme repel hydrophilic substrates. However, substrates that are hydrophobic and contain an opposite charge are strongly attracted.

2. **A** Cofactors are nonprotein substances required for the activity of some enzymes. They include metal ions and coenzymes. Choice B describes coenzymes. Choice C describes prosthetic groups.

3. **B** In enzyme-catalyzed reactions, the reaction rate increases with the addition of increasing substrate until a maximum rate is reached. Then, no increase in reaction rate occurs with the addition of further substrate. Furthermore, the rate of change of the reaction rate falls as the maximum rate is reached. The enzyme has been "saturated" with substrate.

4. **B** Most enzymes show some kind of parabolic change in reaction rate with increasing temperature. High temperatures denature enzymes. Choices A, C, and D are all true statements.

5. **C** Enzymes can have absolute or relative (several substrates can bind enzymes) interactions with substrates. Enzymes catalyze reactions reversibly, by lowering reaction activation energies. The enzyme–substrate bond, however, is noncovalent (hydrogen bonds, ionic bonds, hydrophobic bonds, and van der Waals interactions).

6. **A** Respiration refers to the metabolism of glycolysis products that are broken down to CO_2 and H_2O, using oxygen as a final electron acceptor. Therefore, the process of glycolysis is not included in respiration. The generation of acetyl CoA from pyruvate, the Krebs cycle, and the process of oxidative phosphorylation are all included in respiration.

7. **D** Two net ATP are produced in glycolysis (total of 4 ATP produced, but 2 ATP consumed). The equivalent of 2 ATP (GTP) is produced in the Krebs cycle. A total of 32 ATP are produced by the electron transport and oxidative phosphorylation of NADH and $FADH_2$. Therefore, the percentage of ATP produced by oxidative phosphorylation compared to the total ATP produced by the glycolysis and respiration of glucose is: 32 ATP/36 ATP = 88%.

8. **D** Four ATP are produced directly by glycolysis, but 2 ATP are required to "prime" the system by phosphorylating glucose to glucose-6-phosphate in the first glycolytic step. Two NADH and H^+ are generated in glycolysis, as each of the two three-carbon intermediates from glucose is oxidized.

9. **A** Glycolysis can occur in mammalian cells without mitochondria, because glycolysis occurs in the cytoplasm of cells. Because red blood cells do not have mitochondria, they use glycolytic metabolism extensively for ATP production.

10. **C** The Krebs cycle produces 6 NADH, 2 $FADH_2$, and 2 GTP molecules per glucose molecule metabolized. The question asks for the bioenergetics per mole of acetate. The answer is 3 moles NADH, 1 mole $FADH_2$, and 1 mole GTP.

11. **D** Phosphofructokinase irreversibly phosphorylates fructose-6-phosphate to fructose-1,6-diphosphate. It catalyzes the major rate-limiting step in glycolysis. This enzyme is also the key controlling step of glycolysis. ATP acts as a negative allosteric modulator, whereas ADP acts as a positive allosteric modulator of phosphofructokinase. By this mechanism, the ultimate high-energy product of glycolysis and respiration, ATP, can control the activity of glycolysis.

12. **B** Glycolysis takes place in the cytoplasm, whereas the Krebs cycle occurs in the mitochondrial matrix. Oxidative phosphorylation takes place on the inner mitochondrial membrane.

13. **B** In the Krebs cycle, the electrons removed from the acetyl groups are incorporated into NADH and $FADH_2$.

14. **A** In this difficult question, the two labeled carbons of acetyl CoA add to the four carbons of oxaloacetate (OAA) to give the six-carbon compound, citrate. The two CO_2 molecules that are lost in one turn of the Krebs cycle are not the two carbon units just added by acetyl CoA. They are the carbons that were part of OAA. However, in the subsequent turns of the Krebs cycle, these radioactive carbons are lost.

15. **D** Pyrimidines are single-ring nitrogenous bases. The pyrimidines are cytosine, thymine, and uracil. Purines are two-ring nitrogenous bases; the purines include adenine and guanine.

16. **D** The quantity of G must equal C in DNA because these two bases hydrogen-bond with one another in complementary strands. In addition, the quantity of A must equal T because these two bases hydrogen-bond with one another in complementary strands. Therefore, G = C and A = T, and A + G = C + T. However, A + T ≠ G + C.

17. **A** A glycosidic bond links a base to a pentose sugar. A phosphate ester links a nucleoside to phosphoric acid. Hydrogen bonds bridge two DNA strands.

18. **B** An antiparallel arrangement refers to the inverse orientation of strands. Therefore, if one is given a strand in a 5´ to 3´ orientation, a 3´ to 5´ orientation is antiparallel to it. The question states that the left side of the given DNA fragment (dp) is the 5´ side. Look for an answer choice that is antiparallel to the given DNA sequence. Choice B is complementary to the given DNA sequence.

    The given statement: 5´ ACGGTAC3´

    3´ TGCCATG5´

19. **C** Choice C describes semiconservative replication, the accepted model for DNA replication in eukaryotes. Semiconservative replication produces daughter DNA molecules containing one parental DNA strand paired with one newly synthesized DNA strand. Conservative DNA replication is not the accepted mechanism for DNA replication. In this model, one intact parental and one completely new DNA molecule would result from replication. Choice B is incorrect because bacteria, not mammals, have circular DNA. Choice A is incorrect because it describes conservative replication. Choice D is incorrect because there is no known mechanism by which offspring receive only parental strands.

20. **C** DNA replication in humans is discontinuous. One or both DNA strands may be synthesized in pieces called Okasaki fragments. These fragments are then linked, forming a continuous strand. DNA replication requires several distinct steps:

    1. unwinding of parental DNA (helicase)
    2. creation of RNA primers (primase)
    3. DNA growth in a 5´ to 3´ direction beginning at the 3´ end of the primer (DNA polymerase)
    4. excision of RNA primers (exonuclease)
    5. linking of Okasaki fragments to make a continuous DNA strand (DNA ligase)

21. **D** Primase synthesizes the RNA primer. DNA polymerase adds deoxyribonucleotide bases to the 3´ end of the primer. Exonuclease then removes the primer, DNA polymerase fills in the gaps, and DNA ligase links the remaining Okasaki fragments to make continuous DNA strands.

22. **B** Primase must lay down RNA primers before deoxyribonucleotide bases can be added by DNA polymerase.

23. **B** The best way to understand the similarities and differences between DNA and RNA is to create a table to compare these two molecules.

	RNA	DNA
sugar	ribose	deoxyribose
purines	A, G	A, G
strands	usually single	usually double
pyrimidines	C, U	C, T

24. **A** G–C pairs have three hydrogen bonds, whereas A–T base pairs have two hydrogen bonds. Hydrogen bonds add stability and resistance to denaturation. Therefore, G–C rich DNA is more stable and heat-resistant than A–T DNA.

25. **C** Transcription involves the synthesis of mRNA, rRNA, and tRNA from DNA. It occurs in a 5´ to 3´ direction, and may use either DNA strand. H-bonds between the newly synthesized RNA and parental DNA are weak, allowing dissociation.

HIGH-YIELD REVIEW QUESTIONS

Section I: Molecular and Cellular Biology

Question Set 3

1. Which statement BEST explains why the genetic code is considered degenerate?

   A. One codon can code for several amino acids.
   B. Codons code for amino acids.
   C. Amino acids may form a large variety of possible proteins.
   D. Most amino acids have several codons.

2. Which RNA typically contains the fewest nucleotides?

   A. rRNA
   B. tRNA
   C. mRNA
   D. All types of RNA contain the same number of nucleotides.

3. Which step in translation does NOT consume a high-energy phosphate bond?

   A. Aminoacyl–tRNA (initiator tRNA) binding to the ribosome
   B. Peptidyl–transferase reaction
   C. Amino acid activation
   D. Translocation

4. A peptide is being synthesized on a ribosome. After the growing peptide links a new amino acid to it with a peptide bond, the peptide:

   A. shifts from the "A" site to the "P" site.
   B. shifts from the "P" site to the "A" site.
   C. remains in the "A" site ready to accept a new amino acid.
   D. remains in the "P" site ready to accept a new amino acid.

5. Which type of RNA is NOT required for translation in the cell?

   A. tRNA
   B. rRNA
   C. mRNA
   D. None of the above. All three types of RNA (tRNA, rRNA, and mRNA) are required for translation.

6. Which statement describes the main function of the nucleolus?

   A. The nucleolus directs the transcriptive activities of the nucleus.
   B. The nucleolus coordinates the replication of chromosomal DNA.
   C. The nucleolus synthesizes components of the nuclear membrane.
   D. The nucleolus synthesizes ribosomal RNA.

7. If a portion of a DNA base sequence is G-A-T, the anticodon sequence complementary to the mRNA of this DNA sequence is:

   A. G-A-T.
   B. C-U-A.
   C. G-A-U.
   D. G-U-A.

8. A feature common to the chromosomes of both prokaryotic and eukaryotic cells is the:

   A. presence of DNA and histones in about equal amounts.
   B. circularity of the DNA molecules.
   C. involvement of DNA polymerase in chromosomal replication.
   D. presence of the pyrimidines uracil and cytosine.

9. ATP is a chemical compound classified as a:

   A. nucleoside.
   B. nucleotide.
   C. nucleic acid.
   D. deoxyriboside.

10. *Escherichia coli* is a common intestinal bacterium. One would expect a typical *E. coli* cell to be about the size of a(an).

    A. human liver cell.
    B. polyribosome.
    C. amoeba.
    D. mitochondria.

11. A basic difference between all prokaryotic and all eukaryotic cells is that prokaryotic cells lack a:

    A. cell wall.
    B. plasma membrane.
    C. chromosome.
    D. nuclear envelope.

12. The backbone of a single strand of RNA consists of alternating:

    A. sugars and hydrogens.
    B. hydrogens and nitrogenous bases.
    C. nitrogenous bases and sugars.
    D. none of the above.

13. A cellular organelle found in both mice and bacteria is the:

    A. mitochondrion.
    B. Golgi apparatus.
    C. ribosome.
    D. smooth endoplasmic reticulum.

14. In cellular respiration, the anaerobic steps in glycolysis take place in the:

    A. cytoplasm.
    B. lysosomes.
    C. nucleus.
    D. mitochondria.

15. From which molecule could a cell extract the greatest amount of energy through the normal pathways of aerobic respiration?

    A. $FADH_2$
    B. ATP
    C. NADH
    D. Pyruvic acid

16. Hormones are believed to act at special receptor sites at target cell membrane surfaces to induce an intracellular substance to alter the activity of the target cell. This intracellular substance is known as the "second messenger" of the cell. Which substance is likely to be the intracellular mediator of these hormone functions?

    A. Adenyl cyclase
    B. RNA
    C. Ribonuclease
    D. cAMP

17. Oxidation of molecules is essential for releasing energy in a form that can be used by cells. All cellular oxidations have in common the:

    A. addition of oxygen.
    B. removal of water.
    C. removal of hydrogen.
    D. removal of electrons.

18. Cells are grown in N^{15} for many generations and transferred to N^{14} media. After how many generations does the cellular DNA contain 6.25% N^{15}?

    A. One
    B. Two
    C. Three
    D. Four

19. Which bacteria are rod-shaped?

    A. Bacilli
    B. Spirilla
    C. Cocci
    D. None of the above

20. A certain bacterial strain has a DNA composition of $[G + C] = 80\%$. What percentage of A and U is transcribed from either strand in mRNA? Assume that the DNA coding for the mRNA follows the given composition.

    A. 10%
    B. 20%
    C. 40%
    D. 60%

SOLUTIONS

Molecular and Cellular Biology Set 3

1. **D** The genetic code is considered degenerate because several different codons may code for a single amino acid.

2. **B** The tRNA molecules contain about 75–90 nucleotides, whereas common rRNA's and mRNA's often contain 1000–3000 nucleotides. The rRNA's are fairly large because they make up the structure of ribosomes (along with protein). The mRNA's have many nucleotides because they code for proteins.

3. **B** Only the peptidyl–transferase reaction does not require energy. The aminoacyl–tRNA binding to the ribosome involves initiation factors and energy (GTP). Amino acid activation requires energy (the cleavage of ATP to AMP and PPi). Translocation requires an elongation factor and energy (GTP).

4. **A** New amino acids bound to tRNA come into the "A" site whereas the growing peptide remains in the "P" site. A peptide bond is formed between the amino acid and peptide at the "A" site. The peptide still bound to the tRNA of the last amino acid translocates from the "A" site to the "P" site. This change results in a new codon being positioned at the "A" site ready to be translated.

5. **D** tRNA is required to bring amino acids to the ribosome, with specificity as dictated by the mRNA. rRNA is a critical component of the ribosome. mRNA codes for the peptide to be translated.

6. **D** The nucleolus is involved in the synthesis of ribosomes. It is found in the nucleus of the cell. The nucleolus actively synthesizes ribosomal RNA and helps in ribosome production. With the aid of the nucleolus, an active cell may produce thousands of ribosomes per minute.

7. **C** If the DNA sequence is G-A-T, the mRNA sequence is C-U-A, and the anticodon sequence complementary to the mRNA is G-A-U.

8. **C** Both prokaryotes and eukaryotes require a DNA polymerase to replicate DNA. Eukaryotes have more DNA, presence of histone and nonhistone proteins for DNA packaging, and a nuclear envelope. Prokaryotes often have a single, circular chromosome, less packaging and condensation of DNA, no histones, and no nuclear envelope.

9. **B** ATP (adenosine triphosphate) is a nucleotide. Nucleotides include the nitrogenous base, sugar, and phosphate residue. Nucleosides (e.g., adenosine) include only the base and sugar residue.

10. **D** A bacterial cell is about the size of a mitochondrion. In one popular theory, scientists have postulated that the mitochondria have evolved from bacteria because of the similarities between bacterial and mitochondrial DNA.

11. **D** There are no membrane-bound organelles or nuclear membranes in prokaryotes. Choice A is incorrect because prokaryotic cells (bacteria) have cell walls. Choices B and C are incorrect because prokaryotic cells also have a plasma membrane and a single, circular chromosome.

12. **D** RNA consists of a sugar–phosphate backbone, with the nitrogenous bases guanine, cytosine, uracil, and adenine projecting from the backbone.

13. **C** The ribosome is a non–membrane-bound organelle. It is found in prokaryotes and eukaryotes, although its structure differs somewhat between prokaryotes and eukaryotes.

14. **A** Glycolysis takes place in the cytoplasm of the cell, whereas respiration occurs in mitochondria.

15. **D** Pyruvic acid enters the TCA cycle and leads to the generation of both NADH and $FADH_2$ molecules. Both then enter the oxidative phosphorylation pathway and generate many ATP. A single NADH molecule leads to the production of only 3 ATP. A single $FADH_2$ molecule leads to the production of only 2 ATP.

16. **D** Cyclic AMP (cAMP) is the major intracellular mediator. It modulates the function of many hormones and allows the binding of a cell receptor by a hormone to trigger many intracellular effects.

17. **D** Cellular oxidation refers to the removal of electrons.

18. **D** Solving this problem requires understanding semiconservative replication of DNA. After one generation, only half of the DNA is N^{15} labeled. After a second generation, one fourth of the DNA is N^{15}. By generation three, one eighth is heavy DNA. Finally, by the fourth generation, only one sixteenth of the DNA is N^{15}; that is, approximately 6.25%.

19. **A** Bacilli are rod-shaped, whereas spirilla are corkscrew-shaped and cocci are spherical.

20. **B** If the G + C content of DNA is 80%, the A + T content of the DNA must be 20%. Because A and T of DNA bind to U and A of RNA, respectively, the A and U content of RNA must also be 20%.

HIGH-YIELD REVIEW QUESTIONS

Section II: Physiology

Question Set 4

1. What are the two common portal systems in mammals?
   A. Liver and spleen
   B. Spleen and intestine
   C. Hypothalamus and spleen
   D. Liver and hypothalamus

2. Carbohydrate digestion occurs in many regions of the digestive tract. Carbohydrate digestion first occurs in the:
   A. stomach.
   B. duodenum.
   C. mouth.
   D. small intestine.

3. A unique characteristic of a portal system is that:
   A. it passes through a capillary bed.
   B. veins branch into capillaries.
   C. arteries branch into veins.
   D. it empties into the lymphatic system.

4. During the digestion of carbohydrates, disaccharides are converted to monosaccharides by the addition of:
   A. oxygen.
   B. hydrogen.
   C. NADH
   D. water.

5. Which protein is NOT involved in the digestion or metabolism of carbohydrates?
   A. Amylase
   B. Maltase
   C. Insulin
   D. None of the above; all are involved.

6. The digestive activity in the duodenum occurs:
   A. at acid pH.
   B. once gastric pepsin has digested all proteins to amino acids.
   C. after bicarbonate release by the pancreas.
   D. after secretin release by the pancreas.

7. Movement of contents through the digestive tract of man is characterized by unidirectional oral-to-anal waves of contraction known as:
   A. peristalsis.
   B. posterior progression waves.
   C. pericolonic waves.
   D. regurgitation.

8. The three major segments of the small intestine, in order of passage, are:
   A. duodenum, ileum, jejunum.
   B. duodenum, ascending colon, ileum.
   C. duodenum, jejunum, ileum.
   D. duodenum, ileum, ascending colon.

9. A common medical problem is the blockage of bile flow from the gallbladder to the intestine by a gallstone. One result of this blockage is:
   A. protein malabsorption.
   B. fat malabsorption.
   C. carbohydrate malabsorption.
   D. water-soluble vitamin malabsorption.

10. Gas exchange among individual cells and their immediate environment within the body depends on:
    A. simple diffusion.
    B. osmosis.
    C. facilitated diffusion.
    D. active transport.

11. Which structure has NOT evolved to increase the surface area of an organ or structure?
    A. Alveoli
    B. Villi
    C. Capillaries
    D. None of the above

12. Where does the bulk of gas exchange in human lungs occur?
    A. Trachea
    B. Bronchi
    C. Bronchioles
    D. Alveoli

13. While being swallowed, food is prevented from entering the larynx and progressing into the lungs by apposition of the larynx and the:

    A. epiglottis.
    B. pharynx.
    C. eustachian tube.
    D. esophagus.

14. Which chamber of the heart contains the most highly oxygenated blood?

    A. Left ventricle
    B. Right ventricle
    C. Left atrium
    D. Right atrium

15. Cardiac muscle increases in mass in response to increased resistance or exercise. Realizing that the right and left ventricles are the main pumping compartments, which would be expected to have the thickest wall?

    A. Right
    B. Left
    C. Neither would be as thick as the atria.
    D. They would be of equal thickness because each pumps an equal volume of blood.

16. The signal for contraction of the heart normally originates in which area or region?

    A. Atrioventricular node
    B. Sinoatrial node
    C. Purkinje fibers
    D. His' bundle

17. What does the heart do during the period of systole?

    A. Relaxes and fills
    B. Contracts
    C. Relaxes and then contracts
    D. Pauses awaiting A-V node firing

18. Approximately how much blood does the heart pump per minute in the average human?

    A. 3 liters   C. 7 liters
    B. 5 liters   D. 9 liters

19. Exchange of fluid and nutrients across the capillary wall depends on two carefully balanced forces, namely, hydrostatic and oncotic pressures. If hydrostatic pressure far exceeds oncotic pressure, one would expect:

    A. an increase in intravascular volume.
    B. an increase in interstitial volume.
    C. a decrease in interstitial volume.
    D. A and C.

20. An increase in an intravascular protein, such as albumin, would most likely cause:

    A. an increase in intravascular volume.
    B. an increase in interstitial volume.
    C. a decrease in interstitial volume.
    D. A and C

21. A slight imbalance between oncotic and hydrostatic pressure is present in everyone, leading to slight fluid loss as blood crosses the capillary bed. If this loss goes unchecked, the fluid in the vessels becomes depleted and tissues become swollen. The body counteracts this potential problem by:

    A. decreasing hydrostatic pressure.
    B. increasing oncotic pressure.
    C. resorption into the lymphatics.
    D. increasing venous return to the heart.

22. Which compound is the predominant biologic buffer in the blood plasma?

    A. Phosphate
    B. Bicarbonate
    C. Sulfate
    D. Carbonate

23. The primary functions of the spleen are:

    A. red blood cell storage as well as breakdown and immune function.
    B. synthesis of digestive enzymes and immune function.
    C. filtering of the blood to remove toxins and storing RBCs, breaking down RBCs, or both.
    D. immune function and the production of insulin.

24. Which protein shows positive cooperativity?

    A. Albumin
    B. Hemoglobin
    C. Fibrin
    D. Myoglobin

25. Some organs have multiple sources of arterial blood. One of these organs is the liver. However, the liver has a primary outflow vessel which is (are) known as the:

    A. portal vein.
    B. hepatic veins.
    C. umbilical veins.
    D. hepatic portal vein.

SOLUTIONS

Physiology Set 4

1. **D** The two well-known portal systems in the mammalian body are the liver portal system and the hypothalamic portal system. The liver portal system connects a capillary bed in the intestine to a capillary bed in the liver via the portal vein. The releasing factors produced in the hypothalamus are transported to the anterior pituitary by a portal system.

2. **C** When a food bolus is chewed, saliva mixes with the food. Saliva contains many components, for example, antibodies, antibacterial proteins, and digestive enzymes such as salivary amylase. This enzyme breaks down carbohydrates.

3. **B** A portal system, by definition, consists of two capillary beds connected by a venous network. Therefore, it has an artery–capillary–vein-capillary–vein organization.

4. **D** Hydrolysis of the glycosidic bond between carbohydrate residues requires water.

5. **D** All of these proteins are involved in the digestion or metabolism of carbohydrates. Amylase works on polysaccharides whereas maltase works on the disaccharide maltose. Insulin is required to bring glucose into cells and therefore decreases serum glucose levels.

6. **C** The gastric contents are neutralized by bicarbonate release from the pancreas. Bicarbonate release is stimulated by secretin (a hormone), which is produced by the mucosa of the duodenum.

7. **A** Peristaltic activity is controlled by the autonomic nervous system. Peristalsis propels foodstuffs through the digestive tract. For the other choices given in this question, two are not real terms (posterior progression waves, pericolonic waves) and one is an incorrect term (regurgitation).

8. **C** Choice C gives the correct anatomic relationship. The duodenum leads to the jejunum, then to the ileum. The large intestine follows with its four sections, in consecutive order: ascending colon, transverse colon, descending colon, and sigmoid colon. Finally, the rectum leads to the end of the digestive tract, the anus.

9. **B** Bile emulsifies fats contained in the digestive tract. Once released from the gallbladder under the control of the hormone cholecystokinin, bile enters the small intestine through the common bile duct and mixes with the foodstuffs. Once the fats are emulsified, they can be absorbed in the small intestine.

10. **A** Gas exchange occurs by simple diffusion.

11. **D** The alveoli are spherical and greatly increase the surface area for gas exchange in the lung. Villi and microvilli greatly increase the absorptive surface area in the small intestine. Finally, capillaries greatly increase the surface area for gas exchange, nutrient transfer, thermoregulation, and other functions needed for proper circulatory system function.

12. **D** Although limited gas exchange can occur in the terminal and respiratory bronchioles, by far the majority of gas exchange occurs in the alveoli.

13. **A** The pharynx is the space in the back of the oral cavity, nasal cavity, and throat. The larynx contains the vocal cords. The epiglottis closes against the entryway into the trachea every time swallowing occurs. This act of closing protects the airway from aspiration.

14. **C** The most highly oxygenated blood is found in the heart chamber with blood that has just returned from the lungs, that is, the left atrium. Some oxygen is absorbed by the heart muscle and, therefore, blood entering the left ventricle from the left atrium has slightly less oxygen content.

15. **B** The left side of the heart must pump the blood through the greatest resistance because of the extensive capillary beds of

the systemic circulation. The extra work associated with overcoming this resistance gives rise to the thickness of the left-sided heart muscle.

16. **B** The sinoatrial node (SA node) is the dominant pacemaker of the heart, approximately 100 beats/min in humans. The parasympathetic nervous system has a slowing effect on the SA node, reducing its firing rate to about 70 beats/min. The atrioventricular node is found at the junction of the right atrium and right ventricle and is a backup pacemaker. It beats at an intrinsic rate of about 40–50 beats/min. Purkinje fibers are cells in the heart muscle that also have the capacity to fire spontaneously at a rate of 20–30 beats/min. The general rule is that the center of pacemaking activity with the highest rate dominates the heart rate. The SA node beats the fastest; therefore, it is the dominant pacemaker. The His' *bundle* carries the action potentials from the SA and AV nodes to the right and left ventricles.

17. **B** Systole involves contraction of the heart, whereas diastole involves relaxation of the heart and filling of the chambers.

18. **B** The average human heart pumps approximately 5L of blood per minute at rest.

19. **B** Increased hydrostatic pressure would force fluid out of the vessel and into the interstitial space.

20. **D** Both choices A and C are correct. Increased intravascular protein would increase oncotic pressure. This increase would tend to attract more fluid to stay in the intravascular space and would draw fluid into the vessel from the interstitial space.

21. **C** The lymphatic vessels collect excess fluid that has accumulated in the interstitial space and ultimately return it to the venous circulation.

22. **B** Bicarbonate is the predominant buffer in the plasma.

23. **A** The spleen contains lymphocytes that are important in immune function. The bulk of the spleen is a storage and processing center for old red blood cells that need to be broken down. The pancreas contains the islets of Langerhans, which produce insulin. The pancreas also produces digestive enzymes and bicarbonate. The liver is the dominant organ responsible for filtering the blood and removing toxins.

24. **B** Hemoglobin is the only protein listed that shows positive cooperativity. This term refers to the binding of one oxygen molecule such that each subsequent binding of another oxygen molecule is easier. Hemoglobin has this property because it is a complex multiple-subunit protein. Albumin is a circulating blood protein that binds hormones and drugs. It does not show cooperativity. Fibrin is a clotting protein. Myoglobin is the respiratory pigment in muscle tissue and has only one subunit. It does not show positive cooperativity.

25. **B** The portal vein carries blood from the capillary beds in the intestine to the capillary beds in the liver *without* passing through an arterial vessel; this characteristic qualifies the liver as a portal organ. The hepatic veins carry blood filtered by the liver to the inferior vena cava for return to the heart.

HIGH-YIELD REVIEW QUESTIONS

Section II: Physiology

Question Set 5

1. Increased hydrostatic pressure in Bowman's capsule would tend to:
   A. increase glomerular filtering.
   B. increase reabsorption of amino acids and glucose.
   C. increase reabsorption in the proximal tubule.
   D. decrease glomerular filtering.

2. A drug blocking active transport in the nephron would:
   A. decrease only reabsorption of Na^+.
   B. decrease only reabsorption of Cl^-.
   C. increase the excretion of H^+.
   D. cause the production of isosmotic urine.

3. The hormone aldosterone acts primarily at the:
   A. proximal tubule.
   B. loop of Henle.
   C. distal tubule.
   D. collecting duct.

4. Urine flows into the ureters from the:
   A. urinary bladder.
   B. kidney pelvis.
   C. urethra.
   D. collecting ducts.

5. The white blood cells include:
   A. erythrocytes, basophils, neutrophils.
   B. neutrophils, basophils, eosinophils.
   C. lymphocytes, erythrocytes, basophils.
   D. platelets, lymphocytes, basophils.

6. On a percentage basis, which substance is most completely reabsorbed by the kidneys of a normal person?
   A. Water
   B. Glucose
   C. Urea
   D. Sodium

7. Which description of the autonomic nervous system is NOT true?
   A. It involves the central and peripheral nervous systems.
   B. The parasympathetic division controls all basic visceral functions.
   C. Outflow is via the ventral roots of spinal nerves or through cranial nerves.
   D. It monitors visceral functions over which conscious control is limited.

8. A researcher performs experiments attempting to characterize the action potentials generated by neurons. In terms of the basic properties of action potentials, which finding would be LEAST likely from these experiments?
   A. Action potentials would be found to have a fixed strength.
   B. Action potentials would be found to have a certain maximal rate of generation in nerves.
   C. Action potentials would be found to propagate only unidirectionally in nerves.
   D. Action potentials would be found to cause neurotransmitter release in nerve terminals.

9. Surgical removal of the colon in a human would be expected to cause:
   A. protein insufficiency.
   B. caloric insufficiency.
   C. diarrhea.
   D. constipation.

10. Long bone lengthening occurs:
    A. in the marrow.
    B. in the epiphyseal plate.
    C. in the periosteum.
    D. in the synovial membrane.

11. Which statement about the human reproductive cycle is true?

    A. Negative feedback control maintains nearly constant levels of luteinizing hormone.
    B. Progesterone inhibits release of FSH.
    C. Negative feedback via the nerves from the ovaries to the brain helps control release of FSH by the pituitary.
    D. Ovulation is triggered by external stimulation, such as copulation.

12. Which characteristic is a function of human skin?

    A. Thermoregulation
    B. Emotional expression
    C. Expression of secondary sexual characteristics
    D. All of the above

13. A patient with a pituitary tumor may have polyuria (excessive urine production) as a result of:

    A. decreased angiotensin release.
    B. decreased aldosterone release.
    C. decreased antidiuretic hormone release.
    D. increased antidiuretic hormone release.

14. The formation of a blood clot involves a complex interplay among platelets, clotting factors, and other serum proteins. The final protein involved in the formation of a mature fibrous clot is:

    A. thrombin.
    B. fibrin.
    C. prothrombin.
    D. plasmin.

15. The major lymphatic vessel that empties into the subclavian vein in the thorax is the:

    A. lacteal duct.
    B. common bile duct.
    C. thoracic duct.
    D. lymphatic duct.

16. The human body needs a constant supply of proteins to survive. Therefore, the digestive system is highly efficient in protein digestion and absorption. The first part of the digestive system to begin digesting proteins is the:

    A. mouth.
    B. stomach.
    C. duodenum.
    D. jejunum, ileum, or both.

17. Which hormone increases blood glucose and heart rate?

    A. Insulin
    B. Acetylcholine
    C. Norepinephrine
    D. Cortisone

18. Which gland is derived from ectodermal tissue and contains neuronal tissue?

    A. Adrenal cortex
    B. Thyroid gland
    C. Anterior pituitary
    D. Adrenal medulla

19. The smallest bone in the human body is found in the:

    A. middle ear.
    B. nasopharynx.
    C. hand.
    D. knee.

20. Which nerve tract is the longest?

    A. Parasympathetic, postganglionic
    B. Parasympathetic, preganglionic
    C. Sympathetic, postganglionic
    D. Sympathetic, preganglionic

21. Calcium ions are important in the mechanism of:

    A. nerve terminal function.
    B. muscle contraction.
    C. cardiac muscle action potential.
    D. all of the above.

22. The renin–angiotensin system involves several of the body's organ systems and is important in:

    A. controlling blood pressure.
    B. maintaining buffer systems.
    C. regulating acetylcholine secretion.
    D. regulating blood glucose and fats.

23. Which state of bone is most often associated with osteocytes?

    A. Bone formation
    B. Bone resorption
    C. Bone maintenance
    D. Bone conversion into cartilage

24. Which statement BEST describes the neurological pathways associated with the visual system?
    A. The pathways give rise to visual reflex and interpretive pathways.
    B. The pathways give rise to interpretive pathways associated with the temporal lobes of the cortex.
    C. The pathways give rise to visual reflexes most often associated with the cerebellum.
    D. All of the above are correct.

25. If the dorsal roots of the spinal cord are transected but the ventral roots remain intact, which feature is theoretically LEAST impaired?
    A. Sensation
    B. Monosynaptic reflexes
    C. Controlled movement
    D. All of the above totally impaired

SOLUTIONS

Physiology Set 5

1. **D** Increasing the hydrostatic pressure in Bowman's capsule creates more "back pressure," thereby decreasing the fluid filtering from the glomerulus.

2. **D** Because active transport of ions in the nephron makes water follow passively, blocking active transport mechanisms destroys the nephron's ability to concentrate urine. The urine would then be isosmotic to plasma.

3. **C** Aldosterone, a hormone produced in the adrenal cortex, acts on distal tubular cells of the nephron to increase potassium excretion and decrease sodium loss in the urine.

4. **B** Once urine is formed in the nephron units, it drains via collecting ducts into the renal pyramids and eventually into the renal pelvis. From there, urine enters the ureters and finally the urinary bladder. Urine leaves the body via the urethra.

5. **B** The white blood cells, or leukocytes, include neutrophils, basophils, eosinophils, and lymphocytes. The neutrophils, basophils, and eosinophils are collectively known as granulocytes because of their granular appearance under the microscope. The granulocytes provide nonspecific immunity against foreign invaders such as bacteria and parasites (without the need for antibodies or specific recognition). The lymphocytes provide specific immunity against bacteria, viruses, and tumors.

6. **B** Virtually all the glucose that is freely filtered by the glomerulus is reabsorbed in the proximal convoluted tubule by active transport mechanisms. Although much of the water, urea, and sodium is reabsorbed, all three are also lost to a measurable degree in concentrated urine.

7. **B** The autonomic nervous system has two divisions, the parasympathetic and sympathetic nervous systems. Both send nerve fibers to the visceral organs. For example, the parasympathetic division slows the heart rate whereas the sympathetic division speeds the heart rate. Even in the digestive system, the autonomic nervous system plays an important role. The parasympathetic division increases digestive activity whereas the sympathetic division slßows digestive activity. The other choices are all correct.

8. **C** Action potentials can propagate bidirectionally in nerve tissue. Usually they do not, however, because the region of nerve from which the action potential originates is in a refractory period. The other choices are all true statements.

9. **C** Because the colon is responsible for reabsorbing water from the stool, loss of the colon gives rise to chronic diarrhea. The colon is also a storage place for stool.

10. **B** Long bone lengthens from the bony ends (epiphyses) where the growth plates are located. At these growth plates, or epiphyseal plates, cartilage is converted to bone. Bone width increases when bone grows from the periosteum lining of the bone.

11. **B** Only choice B is true. Once fertilization occurs, the zygote produces human chorionic gonadotropin, which prolongs the corpus luteal production of progesterone. The progesterone is fed back to the hypothalamus to inhibit release of the factors responsible for FSH secretion from the anterior pituitary.

12. **D** All four functions listed are performed by human skin. Remember that the skin has both an epidermis and a dermis. The epidermis is the outer layer of skin, containing rapidly dividing cells and keratin. The dermis contains hair follicles, sweat glands, sebaceous glands, nerve fibers, blood vessels, lymphatics, and other structures.

13. **C** Polyuria would occur as a result of decreased ADH release from the posterior pituitary. ADH (antidiuretic hormone) acts on the collecting tubules

of the nephron to increase water reabsorption. Therefore, decreased ADH leads to less water reabsorption in the kidney, and more urine production.

14. **B** The blood clot is initiated by platelet adherence to a region of injury, usually where collagen is exposed (e.g., a cut or abrasion). The platelet creates a soft clot, or plug, to stop the bleeding. This clot is very fragile. The cascade of proteins that is then activated eventually leads to the conversion of prothrombin to thrombin. Thrombin then converts fibrinogen to fibrin. Fibrin is a fibrous protein that allows cross-linking and solidification of the blood clot.

15. **C** The thoracic duct is the lymphatic vessel that lets the lymphatic fluid from the body reenter the circulatory system. The lymphatic fluid enters the left subclavian vein in the thorax.

16. **C** In the stomach, the digestion of protein begins when the enzyme pepsin, working in the acidic environment of the stomach, begins the proteolytic degradation of protein.

17. **C** Norepinephrine and epinephrine are released by the adrenal medulla at times of acute stress. They increase the heart rate and serum glucose to help the body deal with sudden stress. Norepinephrine is also used by the nervous system as a neurotransmitter. Insulin decreases serum glucose, whereas ADH has no effect on glucose levels. Cortisone, produced by the adrenal cortex, is the hormone of long-term stress. Its action increases serum glucose but has no effect on heart rate.

18. **D** The adrenal medulla is of ectodermal origin and consists of neuron cell bodies.

19. **A** The middle ear bones, the malleus, incus, and stapes, are the smallest bones in the body.

20. **B** The preganglionic parasympathetic nerves are the longest because they originate in the brain and travel all the way to the ganglia, where they synapse. These ganglia are near the target organ. On the other hand, the preganglionic sympathetic nerves originate in the spinal cord and travel only to the sympathetic chain where their ganglia are located. This distance is short. The postganglionic nerves for either the sympathetic or parasympathetic system are shorter than the preganglionic parasympathetic nerves.

21. **D** Calcium ions are important in all muscle functions and in some nerve terminal functions.

22. **A** The renin–angiotensin system works by monitoring blood pressure in the kidney and receives input from the sympathetic nervous system. In situations involving low blood pressure, the kidney releases renin into the circulation. Renin converts angiotensinogen to angiotensin I. Angiotensin I is converted to angiotensin II in the lung. Angiotensin II is a very strong vasoconstrictor.

23. **C** Osteocytes are mature osteoblasts that have become trapped in the formed bone. Osteocytes live in channels of the bone known as *lacunae*. They communicate with other osteocytes and receive nutrition through channels in the bone connecting adjacent lacunae called *canaliculi*. Osteoclasts resorb bone.

24. **A** The visual pathways give rise to pathways that initiate reflexes and enable an individual to interpret images, a function of the occipital region of the cortex. The cerebellum coordinates movement.

25. **C** Controlled movement requires the brain, spinal cord, and ventral roots. Without dorsal roots, no sensation or basic monosynaptic reflexes occur. Movement is probably uncoordinated when sensory input is absent.

HIGH-YIELD REVIEW QUESTIONS

Section II: Physiology

Question Set 6

1. The muscle type lining visceral structures such as the lower esophagus, stomach, blood vessels, and lower gastrointestinal tract is:

   A. voluntary and striated.
   B. involuntary and striated.
   C. voluntary and smooth.
   D. involuntary and smooth.

2. In which region of the spinal cord do afferent (sensory) neurons have their cell bodies?

   A. Ventral root
   B. Dorsal root
   C. Middle region
   D. Interneuronal region

3. Which BEST describes a ganglion?

   A. A proliferation of dense neuronal dendrites
   B. A collection of nerve axons
   C. A collection of cell bodies located in the central nervous system
   D. A collection of cell bodies located in the peripheral nervous system

4. Which primary germ layer gives rise to neurons and brain tissue?

   A. Endoderm
   B. Ectoderm
   C. Mesoderm
   D. None of the above

5. Which hormone is NOT produced by the pituitary gland?

   A. FSH
   B. MSH
   C. Oxytocin
   D. Prolactin

6. The earliest organ system to become functionally active in the human embryo is the:

   A. digestive system.
   B. cardiovascular system.
   C. nervous system.
   D. respiratory system.

7. A researcher is interested in interfering with the development of the heart in an amphibian embryo. Drug A affects the induction of ectoderm from the underlying mesoderm. Drug B blocks the differentiation of mesoderm during gastrulation. Drug C interferes with the invagination process of gastrulation such that the endoderm layer does not form normally. Which drug would BEST accomplish the researcher's goals?

   A. Drug A
   B. Drug B
   C. Drug C
   D. Drug B or C, followed by drug A

8. Which basic germ layer is most often associated with the lining of the respiratory system?

   A. Endoderm
   B. Ectoderm
   C. Mesoderm
   D. None of the above

9. Which germ layer gives rise to the femur and humerus?

   A. Endoderm
   B. Ectoderm
   C. Mesoderm
   D. None of the above

10. Initial cleavage of the zygote produces a cluster of cells. Up until, and including, the time when this cluster reaches the 16-cell stage, it is BEST described as a:

    A. blastula.
    B. morula.
    C. gastrula.
    D. neurula.

11. The proper order of embryonic development is BEST described as:

    A. morula, gastrula, blastula.
    B. blastula, morula, gastrula.
    C. blastula, gastrula, morula.
    D. none of the above.

12. The small yolk found in mammalian development suggests:
    A. polar body formation.
    B. a metabolically active embryo.
    C. a low metabolic requirement of the embryo.
    D. probable placental nutrition.

13. An important pancreatic hormone that promotes glucose uptake by muscle and fat cells is:
    A. insulin.
    B. glucagon.
    C. secretin.
    D. cortisol.

14. Which endocrine cells produce pancreatic glucagon?
    A. Alpha cells
    B. Beta cells
    C. Gamma cells
    D. Delta cells

15. A diabetic patient has given himself an accidental overdose of insulin. What is the most logical treatment for this patient considering the effect of excess insulin?
    A. Give the patient intravenous insulin.
    B. Give the patient an intravenous drug that blocks the effect of glucagon.
    C. Give the patient a drug that stimulates the beta cells in the pancreas.
    D. Give the patient intravenous glucose.

16. The role of parathyroid hormone is BEST described as:
    A. increasing serum phosphate concentration.
    B. decreasing serum calcium concentration.
    C. increasing serum calcium concentration.
    D. decreasing serum sodium concentration.

17. Calcitonin is a hormone that decreases serum calcium. If the decision is to interfere with the gland that produces calcitonin, which gland should be targeted?
    A. Parathyroid gland
    B. Pituitary gland
    C. Thymus gland
    D. Thyroid gland

18. Which pair of hormones is produced by the adrenal medulla?
    A. Cortisone and hydrocortisone
    B. Aldosterone and androgens
    C. FSH and LH
    D. Epinephrine and norepinephrine

19. The hormones of the adrenal cortex are all:
    A. glycolipids.
    B. proteins.
    C. peptides.
    D. steroids.

20. Which hormone is transported from the hypothalamus to the posterior pituitary by axonal transport?
    A. FSH
    B. Prolactin
    C. ADH
    D. MSH

21. All of the following are characteristic of chordates EXCEPT:
    A. a ventral nerve cord.
    B. segmented body musculature.
    C. a ventral heart.
    D. a notocord.

22. The unit of contraction of striated muscle has been studied extensively. This structure, which is bound by two Z lines and contains actin, myosin, troponin, and tropomyosin, is known as the:
    A. sarcolemma.
    B. sarcoplasm.
    C. sarcosome.
    D. sarcomere.

23. Which cell type is the most highly differentiated?
    A. A blastula cell
    B. An intestinal epithelial cell
    C. A stem cell in the bone marrow
    D. An ectodermal cell in the retina of the developing eye of an embryo

24. A person is very energetic, perspires profusely, loses weight, and has bulging eyes. Which endocrine gland malfunctions and causes these conditions?
    A. Adrenal cortex
    B. Posterior pituitary
    C. Parathyroid
    D. Thyroid

25. Impulses traveling through the nervous system are transmitted electrically in neurons and chemically across synapses. Well-known synaptic neurotransmitters do NOT include:

A. acetylcholine.
B. dopamine.
C. serotonin.
D. any of the above, because all of the choices are well-known synaptic neurotransmitters.

SOLUTIONS

Physiology Set 6

1. **D** The muscle that lines the hollow organs of the body is smooth muscle. This muscle type is characterized by slow contractions, spindle-shaped cells, involuntary control, and innervation by the autonomic nervous system.

2. **B** The cell bodies for afferent sensory nerves are located in the dorsal root ganglion. The ventral root is where motor nerves exit. Interneurons synapse in the spinal cord proper.

3. **D** Several terms are used in biology to describe nerves and collections of nerve cell bodies. A ganglion is a collection of nerve cell bodies found in the peripheral nervous system. A collection of nerve cell bodies in the central nervous system is known as a nucleus. A collection of nerve axons is known as a nerve.

4. **B** Ectoderm gives rise to the nervous system, skin, skin appendages such as hair and nails, the cornea of the eye, the enamel of the teeth, and so forth.

5. **C** The hypothalamus produces oxytocin and ADH and transports them axonally to the posterior pituitary for release. The other hormones listed are produced in, and released by, the anterior pituitary.

6. **C** The nervous system is the earliest organ system to form and function in the human embryo.

7. **B** Drug B blocks the formation and differentiation of mesoderm. Because the heart and cardiovascular system are derived from mesoderm, drug B has the greatest effect. It is not possible to know whether any of the other drugs would have an effect on the heart, so B is the best choice.

8. **A** The lining of the hollow organs, such as the lungs and digestive organs, is endodermal in origin.

9. **C** Bone tissue is derived from mesoderm.

10. **B** A morula is a solid ball of cells derived from the early cleavage events of the zygote. Once the morula reaches the 16-cell stage, fluid is secreted into the central region, and a hollow ball of cells known as a blastula forms.

11. **D** The correct order of development is morula, blastula, gastrula.

12. **D** The placenta, a mammalian organ consisting of fetal and maternal components, aids in the exchange of materials between the fetus and the mother. The presence of a placenta means that outside nutrient requirements in the form of yolk are minimal. Therefore, mammals have little yolk.

13. **A** Glucagon releases glucose from the glycogen stores in the liver and muscle. Insulin, on the other hand, decreases serum glucose by promoting muscle and fat-cell glucose uptake.

14. **A** Pancreatic glucagon is secreted by the alpha cells. The beta cells produce insulin, whereas the delta cells produce somatostatin.

15. **D** With an overdose of insulin, blood sugar is very low because insulin is required to bring glucose into many cells of the body. Therefore, the available serum glucose is brought into the cells. Replacement with intravenous glucose is most logical. Glucagon should not be blocked because it acts to increase serum glucose. In addition, choice A would be disastrous, further decreasing serum glucose and likely leading to death. Choice C is incorrect because beta cells produce insulin, which further decreases serum glucose.

16. **C** Parathyroid hormone increases serum calcium by increasing bone resorption and decreasing losses of calcium in the urine. Serum phosphate decreases in the presence of parathyroid hormone.

17. **D** Calcitonin is produced by the thyroid gland, and parathyroid hormone is produced by the parathyroid gland. Thyroid hormone is also produced by the thyroid gland.

18. **D** Cortisone is produced by the adrenal cortex, as are aldosterone and androgens. Epinephrine and norepinephrine are produced by the adrenal medulla.

19. **D** Adrenocortical hormones are steroids, whereas the adrenal medullary hormones are aromatic ring-containing amines. The pituitary hormones are small glycopeptides.

20. **C** The hormones ADH and oxytocin are transported from the hypothalamus to the posterior pituitary by axonal transport.

21. **A** Chordates are characterized by a *dorsal nerve cord*, a segmented body musculature, a ventral heart, a notocord, and pharyngeal gill slits sometime during development.

22. **D** The sarcomere is the unit of function of muscle tissue. It is the repeating unit of striated muscle. The borders of a sarcomere are delineated by Z-lines. The thin filaments, containing actin, bind to the Z-lines. The thick filaments, containing myosin, troponin, and tropomyosin, are found within the sarcomere. The sarcoplasmic reticulum is a modified form of endoplasmic reticulum that is found in muscle cells and stores calcium. This calcium is released to trigger muscle contraction.

23. **B** A highly differentiated cell type has undergone development changes from an immature to a mature, committed form. An intestinal epithelial cell cannot develop any further; it is committed to a specific function. The other choices describe cells that have yet to become fully differentiated.

24. **D** Excessive levels of thyroid hormone pathologically increase metabolic rate and may cause the other symptoms described.

25. **D** The common neurotransmitters that operate across synapses include acetylcholine, norepinephrine, dopamine, and serotonin.

HIGH-YIELD REVIEW QUESTIONS

Section II: Physiology

Question Set 7

1. Which hormone is NOT secreted by the anterior pituitary gland?
   A. FSH
   B. GH
   C. ADH
   D. ACTH

2. The usual result of moderate parasympathetic vagus nerve stimulation is an:
   A. increased breathing rate.
   B. increased peristaltic rate.
   C. increased heart rate.
   D. irregular heart beat.

3. In humans, the highly coiled duct in which most of the sperm are stored before ejaculation is called the:
   A. epididymis.
   B. seminal vesicle.
   C. vas deferens.
   D. urethra.

4. Oxygenated blood is carried to the heart from the lungs via:
   A. coronary veins.
   B. coronary arteries.
   C. pulmonary veins.
   D. pulmonary arteries.

5. The portion of the brain that coordinates muscular activity is the:
   A. corpus callosum.
   B. cerebellum.
   C. cerebrum.
   D. medulla oblongata.

6. During meiosis, each primary spermatocyte gives rise to:
   A. two sperm.
   B. two sperm and two polar bodies.
   C. two polar bodies.
   D. four sperm.

7. A person of blood type A can:
   A. be the parent of a child with blood type B.
   B. possess only the B cellular antigen.
   C. possess both the A and B cellular antigens.
   D. safely receive blood from a donor of type B.

8. The postovulatory follicle that actively produces progesterone is called the:
   A. corpus luteum.
   B. corpus albicans.
   C. corpus striatum.
   D. corpus folliculus.

9. Which organ, or specific area in an organ, has major control of the pH homeostatic mechanism in man?
   A. Liver
   B. Kidney
   C. Intestine
   D. Medulla oblongata

10. All of the following glands secrete digestive enzymes EXCEPT the:
    A. pancreas.
    B. salivary glands.
    C. stomach.
    D. liver.

11. The blood-flow sequence after absorption of nutrients in the small intestine is:
    A. liver, hepatic vein, hepatic portal vein, inferior vena cava, right atrium.
    B. hepatic vein, liver, hepatic portal vein, inferior vena cava, right atrium.
    C. hepatic portal vein, liver, hepatic vein, inferior vena cava, right atrium.
    D. none of the above.

12. In a given muscle, it is observed that the strength of muscle contraction is increasing. The BEST explanation for stronger muscle contractions is:

A. increased strength of action potentials.
B. more action potentials acting at the dorsal root ganglion.
C. alternate activity of flexor–extensor muscle antagonists.
D. increased motor neuron activation of motor units.

13. Which choice is NOT a product of the mammalian liver?

A. Urea
B. Bile
C. Cholecystokinin
D. Cholesterol

14. Mammalian somatic motor neurons:

A. leave the spinal cord via the ventral root.
B. enter the spinal cord via the dorsal root.
C. innervate pressure receptors in muscle.
D. innervate visceral organs.

15. In the U-shaped tube with a semipermeable membrane separating solution A from solution B, what happens to the water level?

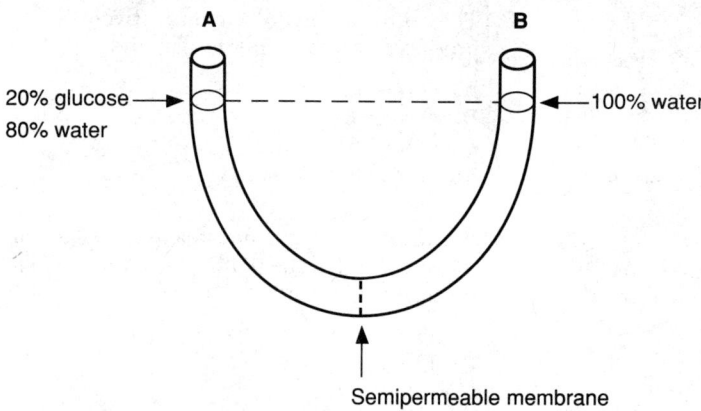

A. It rises inside A because water passes from the area of greater concentration to the area of lesser concentration of water.
B. It rises inside B because water passes from the area of lesser concentration to the area of greater concentration of water.
C. It remains the same because atmospheric pressure is equal on both sides of the system.
D. Not enough information is present to predict the outcome.

16. In the human fetal circulatory system, which blood vessel carries the most oxygenated blood?

A. Ductus arteriosus
B. Pulmonary vein
C. Umbilical artery
D. Umbilical vein

17. The hormone oxytocin is secreted from the:

A. medulla.
B. anterior pituitary.
C. posterior pituitary.
D. adrenal gland.

18. Which hormone leads to a decrease in blood glucose?

A. Cortisol
B. Glucagon
C. Thyroid hormone
D. Oxytocin

19. A drug that stimulates the sympathetic nervous system is administered to a patient. This drug can be expected to:

A. slow the heart rate.
B. increase intestinal peristalsis.
C. constrict the bronchial tubes.
D. dilate the pupils.

20. Failure of separation of the cross-bridges from the actin filaments in muscle tissue is known as:

    A. tetanus.
    B. fasciculation.
    C. isotonic contraction.
    D. rigor mortis.

21. Soon after fertilization, the dividing zygote forms a hollow ball of cells surrounding a central cavity. Which stage is described?

    A. Morula
    B. Blastula
    C. Gastrula
    D. Neurula

22. In mammalian embryonic development, the embryo proper develops from the:

    A. trophoblast.
    B. amnion.
    C. inner cell mass.
    D. primary yolk sac.

23. Which morphogenetic movement forms the gastrula of vertebrate embryos?

    A. Differentiation
    B. Cleavage
    C. Determination
    D. Invagination

24. Which tissue is NOT one of the extraembryonic membranes of the human embryo?

    A. Amnion
    B. Allantois
    C. Yolk sac
    D. None of the above

25. A biologist conducts an experiment in which the dorsal lip of the blastopore from an early frog embryo is inserted under the ectoderm on the ventral surface of a second early embryo. What probably happens to the host embryo?

    A. It dies in reaction to the foreign tissue.
    B. It develops a second primitive streak.
    C. It develops a second neural groove.
    D. It develops an extra pair of legs.

SOLUTIONS

Physiology Set 7

1. **C** ADH is produced in the hypothalamus and is moved by axonal transport to the posterior pituitary for release. The other hormones listed are released by, and produced in, the anterior pituitary. The anterior pituitary produces hormones under the direction of the hypothalamus. The hypothalamus produces releasing factors, which reach the anterior pituitary through a portal blood circulation and signal the anterior pituitary to produce hormones. It is important to recognize that unlike the anterior pituitary, the posterior pituitary gland does not produce hormones.

2. **B** The parasympathetic nervous system plays a major role in activating digestive activity, whereas the sympathetic nervous system slows digestive activity. Mnemonics is a good way to remember many of the functions of the autonomic nervous system. The sympathetic nervous system is associated with the "fight or flight" reaction, whereas the parasympathetic nervous system is associated with the "rest and ruminate" reaction. In stressful situations, such as fight or flight, the sympathetic nervous system acts to increase blood circulation and oxygen flow to the central core of the body and to the muscles. This action causes the heart rate and respiratory rate to increase, the bronchial tubes to dilate, the digestive activity to slow down, the skin blood vessels to vasoconstrict, and so forth. In restful situations, the parasympathetic nervous system is active, and this situation causes increased digestive activity, lower heart rates, narrowing of bronchial tubes, lower respiratory rates, and so forth.

3. **A** The seminal vesicles, prostate, and Cowper's glands produce semen. Sperm are stored and mature in the epididymis.

4. **C** The pulmonary veins carry blood from the lungs to the left side of the heart. This blood is the most highly oxygenated, because no oxygen has yet been absorbed from it.

5. **B** The cerebellum coordinates muscular activity. The corpus callosum is a region of white matter that connects the two cerebral hemispheres. The cerebrum is the cerebral cortex, the region of higher function, such as thinking and memory. The hypothalamus controls basic drives and emotions such as sex, thirst, hunger, and temperature regulation. The medulla is a center for basic life support, for example, breathing and heart beat control.

6. **D** Four sperm and no polar bodies are produced from spermatogenesis. These sperm are all haploid and of equal size. These sperm are all equally able to fertilize an egg cell. Note the difference between spermatogenesis and oogenesis. In oogenesis, only one viable haploid oocyte is produced.

7. **A** A person with type A blood has type A antigens on their red blood cells. That person does not have type B antigens; however, the person does have antibodies against other blood antigens, such as type B. A person with type O blood is a universal donor and has no antigens. A person with type AB blood is a universal recipient and has both A and B antigens, but no antibodies against either A or B.

8. **A** The corpus luteum forms after the oocyte leaves the graafian follicle. It secretes progesterone and eventually involutes to become the corpus albicans if fertilization does not occur.

9. **B** Both the kidney and lung are important in acid–base balance. The lung controls more minute-to-minute changes, whereas the kidney controls pH over the longer term. The lung is able to control more rapid changes in blood pH through the bicarbonate or carbonic acid buffer system or both. The kidney controls long-term regulation of blood pH by the absorption or secretion of bicarbonate ion and hydrogen ion, or both.

10. **D** The liver secretes no digestive enzymes. All the other glands do, including the

salivary glands (salivary amylase). The stomach secretes the digestive enzyme pepsin. The pancreas secretes many digestive enzymes, including trypsin, chymotrypsin, amylase, lipase, DNAase, and RNAase.

11. **C** Once nutrients are absorbed in intestinal capillaries, they travel by the portal vein to the liver. Once filtered through the liver, these nutrients move through the hepatic veins to the inferior vena cava to the heart.

12. **D** The way to increase the strength of contraction is to increase the activation of motor units. This increased activation recruits more sarcomeres and therefore increases contraction strength. Choice A is incorrect because action potentials are generally all-or-none phenomena. Although it is possible to affect the peak of an action potential (by altering the concentration of sodium and potassium ions in solution), this change does not result in a stronger muscle contraction. Choice B is incorrect because the dorsal root ganglion is a sensory ganglion. It is the place where the cell bodies of sensory neurons entering the spinal cord are located. Choice C is incorrect because alternate activity of flexor muscles and extensor muscles leads to flexion and extension movements or movements that cancel out one another. The alternate activity of flexors and extensors does not lead to stronger muscle contractions.

13. **C** CCK, or cholecystokinin, is produced in the duodenum and acts on the gallbladder to cause the release of stored bile. The other choices are all produced in the liver.

14. **A** Motor neurons leave the cord via the ventral roots, whereas sensory input reaches the cord via the dorsal roots.

15. **A** Although the concentration of water is not generally discussed, choice A is correct. Water moves so as to dilute concentrated solutes and the level rises on side A.

16. **D** The umbilical vein carries blood just oxygenated in the placenta to the fetus. The other vessels carry blood that is less oxygenated. The least oxygenated blood is carried by the umbilical artery because this vessel carries deoxygenated blood from the fetus toward the placenta.

17. **C** Oxytocin is secreted from the posterior pituitary. Recall that this hormone, along with ADH, is produced in the hypothalamus and transported to the posterior pituitary. It is secreted from the posterior pituitary, and acts to cause smooth muscle contraction in the breasts (during lactation) and uterus (during labor).

18. **C** Thyroid hormone increases metabolism and therefore leads to the use of blood glucose. Cortisol and glucagon both increase blood glucose.

19. **D** Sympathetic stimulation dilates the pupils (seen also in cocaine toxicity), speeds heart rate, increases breathing rate, slows digestion, and so forth.

20. **D** Rigor mortis occurs when cross-bridges form between the myosin heads and actin and do not release. Remember that separation of the myosin heads from actin requires ATP. Once death occurs, no ATP is available and rigor mortis sets in.

21. **B** The blastula stage follows the morula stage of development. A blastula is a hollow ball of cells that are usually undetermined.

22. **C** Once it reaches the 16-cell stage, the zygote is known as a morula. The morula consists of a group of centrally located cells, the inner cell mass. This mass gives rise to the embryo proper. The surrounding outer cell layer, the outer cell mass, later forms the trophoblast, which is the fetal contribution to the placenta.

23. **D** Invagination converts the blastula to the gastrula, and allows the differentiation of three germ layers.

24. **D** The amnion, allantois, and yolk sac are extraembryonic tissues from the embryo.

25. **C** The dorsal lip of the blastopore leads to the differentiation of a neural groove. Transplantation of the dorsal lip early in development gives rise to an extra neural groove if the dorsal lip is placed beneath undifferentiated ectoderm. This situation is a classic example of tissue induction.

HIGH-YIELD REVIEW QUESTIONS

Section III: Genetics, Evolution, and Botany

Question Set 8

1. A researcher crossbreeds purebred tall plants and purebred dwarf plants. All the offspring of these crosses are tall. These offspring are then crossed. One can predict that in the F_2 generation:

   A. 1/4 of the offspring will be tall.
   B. 1/2 of the offspring will be tall.
   C. 3/4 of the offspring will be tall.
   D. all the offspring will be tall.

2. In question 1, the allele for the dwarf plants:

   A. is recessive in both the F_1 and F_2 generations.
   B. is recessive in the F_1 generation.
   C. is absent in the F_1 generation.
   D. is absent in the F_1 or F_2 generation.

Questions 3–5 refer to the following data:

Genes for dark eyes (D) are usually dominant over genes for blue or gray eyes (d). A man with dark-brown eyes marries a woman with light-gray eyes. They have two children, a boy with dark-brown eyes and a girl with blue eyes.

3. The genotype of the man is:

   A. DD.
   B. Dd.
   C. dd.
   D. unable to be determined from the information.

4. The genotype of the wife is:

   A. DD.
   B. Dd.
   C. dd.
   D. unable to be determined from the information.

5. The genotype of the son is:

   A. DD.
   B. Dd.
   C. dd.
   D. unable to be determined from the information.

6. A man with brown eyes marries a woman with blue eyes. They have 12 brown-eyed children. What are the probable genotypes of the man, his wife, and all the children?

   A. BB, Bb, bb
   B. Bb, bb, Bb
   C. BB, Bb, Bb
   D. None of the above

7. Genes occupying homologous loci on a pair of homologous chromosomes are called:

   A. linked.
   B. segregated.
   C. recombined.
   D. alleles.

8. In humans, albinism is determined by a recessive gene (a). Syndactyly, a condition in which two or more fingers or toes are joined by a web of muscle and skin, is determined by a dominant gene (B). If a couple knows that their genotypes are BbAa and bbAa, what is the probability of both their offspring being albino and having joined fingers or toes?

   A. 1/8
   B. 1/4
   C. 1/2
   D. 1/6

9. After karyotypic analysis, a patient is determined to have 22 pairs of autosomes and two pairs of X chromosomes. This patient is:

   I. phenotypically female.
   II. genotypically normal.
   III. phenotypically male.
   IV. genotypically abnormal.

   A. I and II
   B. II and III
   C. I and IV
   D. III and IV

10. Sometimes in meiosis, one of the homologous pairs of chromosomes fails to separate properly and both chromosomes move to the same pole. This occurrence is known as:

   A. linkage.
   B. crossing-over.
   C. chiasma.
   D. nondisjunction.

11. In humans, an odd number of chromosomes in a cell can indicate:

   A. an abnormal cell.
   B. a gamete.
   C. a haploid cell.
   D. all of the above.

12. In a dihybrid cross, the effect of crossing one pair of factors is independent of the effect of crossing the second pair of factors. The principle exemplified is termed the:

   A. law of segregation.
   B. law of independent assortment.
   C. law of nondisjunction.
   D. law of linkage.

13. A sex-linked recessive trait carried by the X chromosomes rarely shows the trait phenotypically in females because:

   A. only one X chromosome is functional in most female cells.
   B. the female must inherit two recessive alleles, one from each gamete.
   C. the recessive allele is normally linked with the dominant one for females.
   D. recessive phenotypes rarely occur in females.

14. If a strain of pure-breeding red flowers is crossed with a strain of pure-breeding white flowers, all the F_1 offspring are pink. A probable explanation is that:

   A. a mechanism of codominance is responsible.
   B. a mechanism of incomplete dominance is responsible.
   C. a mechanism of linkage is responsible.
   D. a mechanism of independent assortment is responsible.

15. Assuming that color blindness is a sex-linked recessive trait, the offspring of a normal man and a color-blind woman are:

   A. all color-blind sons and half carrier daughters.
   B. half color-blind sons and half carrier daughters.
   C. half color-blind sons and all carrier daughters.
   D. all color-blind sons and all carrier daughters.

16. A germ-cell chromosome contains the alleles A and B, and its homologous chomosome contains the alleles A´ and B´. Two gametes derived from this germ-cell line contain chromosomes with the AB´ and A´B genotype. The process that can account for this configuration is:

   A. mutation.
   B. crossing-over.
   C. recombination.
   D. independent assortment.

17. Replication of DNA during cell division occurs in:

   A. interphase.
   B. prophase.
   C. metaphase.
   D. anaphase.

18. During which phase of mitosis do chromosomes line up along the equator of the cell?

   A. Interphase.
   B. Prophase.
   C. Metaphase.
   D. Anaphase.

19. During cell division, the cell component that moves to the opposite pole of the cell preceding the movement of the chromosome is the:

   A. centromere.
   B. centriole.
   C. sister chromatid.
   D. centrosome.

20. Meiosis of ova usually involves:

   A. equal divisions with at least two functional cells.
   B. equal divisions with only one functional cell.
   C. unequal divisions with at least two functional cells.
   D. unequal divisions with only one functional cell.

21. The mitotic period during which the nuclear membrane and nucleoli disappear is:

    A. interphase.
    B. prophase.
    C. metaphase.
    D. anaphase.

22. The nucleus of a somatic cell of *Drosophila* contains eight chromosomes. During which stage of cell division would a cell of *Drosophila* have 16 chromosomes?

    A. Metaphase of meiosis II
    B. Telophase of mitosis
    C. Prophase of meiosis II
    D. Anaphase of mitosis

23. The genotype of an individual showing a particular dominant trait can be determined by:

    A. crossbreeding.
    B. backcrossing.
    C. outbreeding.
    D. karyotypic analysis.

24. In humans, the major blood-type differences are determined by multiple alleles designated I^A, I^B, and i. I^A and I^B are codominants and i is recessive. Gene combinations $I^A I^A$ and $I^A i$ produce type A blood. $I^A I^B$ produces type AB blood; $I^B I^B$ and $I^B i$ produce type B blood; and ii produces type O blood. If one parent has type A blood, the other has type B blood, and they have a child with type O blood, what is the probability of their next child having type AB blood?

    A. 0%
    B. 25%
    C. 50%
    D. 100%

25. The following diagram shows a chromosomal map with several allelic loci. Which two loci can be expected to give the highest crossover or recombination frequency?

    ```
    A    B    C    D    E
    ─────────────────────

    A'   B'   C'   D'   E'
    ```

    A. A and C´
    B. B´ and E
    C. E´ and D
    D. A´ and E

26. How many different gametes can be produced from an organism with the genotype CCDdEEFfgg?

    A. 16   C. 6
    B. 8   D. 4

27. Approximately what day after the start of menstruation does human ovulation occur?

    A. Day 1   C. Day 8
    B. Day 5   D. Day 14

28. The following diagram shows a human pedigree. Squares represent males; circles represent females. Shaded symbols indicate people with a certain trait; empty symbols indicate absence of the trait. Based on common inheritance patterns and the following pedigree, inheritance is most likely:

    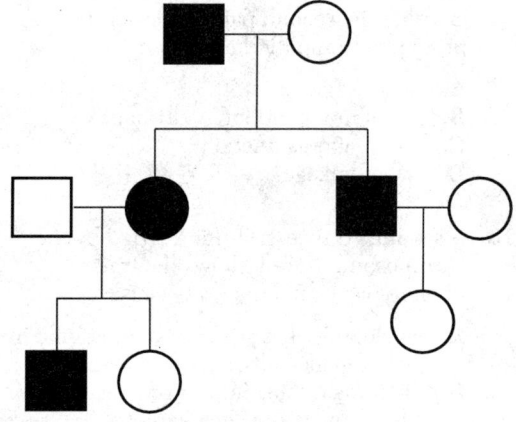

    A. sex-linked dominant.
    B. sex-linked recessive.
    C. autosomal dominant.
    D. autosomal recessive.

29. The frequency of blue eyes in a population is 0.2. Assume that all other members of the population have brown eyes. If brown eyes are dominant over blue eyes, what is the frequency of brown-eyed heterozygotes in this population? Assume H–W equilibrium.

    A. 0.16   C. 0.08
    B. 0.64   D. 0.32

30. Color blindness is a sex-linked trait. What genotype is most likely for the father of a color-blind girl?

    A. $X^c Y^c$ C. XY^c
    B. $X^c Y$ D. XY

SOLUTIONS

Genetics, Evolution, and Botany Set 8

1. **C** This question describes monohybrid crosses. For the first cross described, use the following cross: TT × tt = all Tf offspring (F_1 generation). For the second cross, Tt × Tt = 1 TT, 2 Tt, and 1 tt (F_2 generation). Therefore, 3/4 offspring are of the tall phenotype.

2. **A** The allele for dwarf plants must be recessive to give phenotypically tall plants in the observed ratios in both the F_1 and F_2 generations.

3. **B** With such problems, think about what parental genotypes can satisfy the genotypes generated by the crosses. If the cross assumes that the father is Dd and the mother is dd, the offspring are Dd or dd: Dd × dd = Dd or dd offspring. Therefore, the father is Dd.

4. **C** The explanation for question 3 shows that the mother has a dd genotype.

5. **B** See the solution for question 3. Notice that the sex of the children is irrelevant to this problem. Only the genotype of the nonsex chromosomes is important here.

6. **D** Because brown eyes are dominant over blue eyes and all the children have brown eyes, the father is probably homozygous (BB) and the mother is probably homozygous (bb) recessive. The children are probably all heterozygous (Bb).

7. **D** Alleles are alternative forms of a gene. They are found occupying similar loci on pairs of homologous chromosomes.

8. **A** The chance of a child being albino equals the chance of a child inheriting the (aa) genotype. From each parent comes a 1/2 chance of inheriting an albino allele. Therefore, the chance of inheriting the (aa) genotype = 1/2 × 1/2 = 1/4. The chance of inheriting syndactyly equals the chance of inheriting a single B allele (the allele is dominant; only one allele is needed). Only one parent has a B allele, and the chance of inheriting it is 1/2. To inherit both syndactyly and albinism: 1/4 × 1/2 = 1/8.

9. **C** Lack of the Y chromosome causes the female phenotype. Normal females have 22 pairs of autosomal (nonsex) chromosomes and one pair of X (sex) chromosomes.

10. **D** The question describes nondisjunction. In this process, the homologous chromosomes do not separate properly in meiosis. This process may give rise to gametes with abnormal numbers of chromosomes. Choices A, B, and C describe different genetic terms. A chiasma is a region that is visible microscopically (appears X shaped under the microscope) and represents the crossing-over of two chromatids belonging to separate, but homologous chromosomes. Crossing-over is the process in which genetic material is exchanged between homologous chromosomes in prophase I of meiosis. Linkage implies genes that are inherited as units. For example, genes located on the same chromosome are said to be linked because they are inherited together in genetic crosses. Sex-linkage has a different meaning; it implies that a gene or genes are linked to a sex chromosome (either X or Y). These sex-linked genes are passed to offspring with either the X or Y chromosome.

11. **D** Nondisjunction during meiosis can cause an odd number of chromosomes in gametes and, therefore, either 45 or 47 chromosomes in diploid cells (instead of the normal 46). Gametes normally have 23 chromosomes each. When two gametes fuse, the normal diploid compliment of 46 chromosomes is achieved.

12. **B** The law of independent assortment states that during gamete formation, alleles for different genes segregate independently. Therefore, in dihybrid crosses, the genes are independently packaged in different combinations of alleles based on probabilities (classic 9:3:3:1) ratio. Choice A is incorrect because the law of segregation describes the sorting of alleles into separate gametes, not the crossing and sorting of one pair of alleles independent of a second set of alleles. Choices C and D are incorrect and are described in the solution to question 10.

13. **B** In sex-linked diseases, females rarely show the condition because females must inherit two X chromosomes with the defective gene. If the female has one X chromosome with the gene and one normal chromosome, she is a carrier of the disease. A male needs only one X chromosome with the defective gene to be afflicted with the disease. Males cannot be carriers of X-linked traits.

14. **B** Incomplete dominance allows a "blending" of the genes. Therefore, the progeny of a cross showing incomplete dominance may have a phenotype that is a combination of the parental phenotypes. Choice A is incorrect because codominance produces a phenotype in which both alleles are expressed in heterozygotes. Choices C and D are incorrect because linkage and independent assortment cannot explain the production of pink flowers from red and white parental phenotypes ("blending" of phenotypes).

15. **D** To solve this question, set up the described cross and evaluate the offspring produced by the cross. $XY \times X^cX^c = X^cY$ (color-blind sons) + XX^c (carrier daughters).

16. **B** Crossing-over occurs during prophase of meiosis I and gives rise to diversity in the gene pool. Choice A is incorrect because mutations are rare and random events. It is extremely unlikely that the gametes have obtained their new genotypes by mutation. Choices C and D are incorrect because these terms do not describe processes that can produce the allelic changes given in this question. The solution to question 12 describes the process of independent assortment. Recombination occurs when offspring of crosses have different combinations of phenotypes from the parental generation, simply owing to the independent assortment of alleles that occurs in meiosis.

17. **A** DNA replication occurs during interphase.

18. **C** During the metaphase stage of mitosis, the chromosomes line up along the equator of the cell (equitorial plate) in preparation for separation.

19. **B** The centriole helps organize microtubule assembly during cell division by moving to the opposite end of the cell before chromosome movement. The centromere is the central region that joins two sister chromatids. Sister chromatids are the identical copies of a replicated chromosome in a cell. During DNA replication in interphase, two identical copies are made of the information present in a chromosome. The identical copies of the genetic information present in a single chromosome are called sister chromatids. Centrosome is a nonsense term.

20. **D** An unequal distribution of cytoplasm occurs in an effort to produce a single oocyte with a large quantity of yolk. Three polar bodies are produced in the process.

21. **B** During prophase of mitosis, the nuclear membrane and nucleolus are no longer visible with light microscopy.

22. **D** DNA replicates in interphase. Chromosome condensation occurs in prophase, and separation of these double-stranded chromosomes (i.e., those with two chromatids) into new single-stranded chromosomes (one of the identical sister chromatids) occurs during anaphase. Therefore, 8 "chromosomes" become 16 "chromosomes" before cytokinesis is complete.

23. **B** The backcrossing technique can be used to determine the genotype of an individual. Crossbreeding implies that breeding occurs across different strains or variants of a species. An example of crossbreeding is mating a poodle with a Great Dane. Outbreeding is a process that selectively attempts to eliminate a particular trait or traits from a species by selectively mating individuals that do not have the trait or show a mild form of the trait.

24. **B** The fact that the child is type O (the ii genotype) means that the parents must be I^Ai and I^Bi. Considering the genotypes of the parents, the chance of getting a I^AI^B child is 1/4, or 25%.

25. **D** The highest frequency of recombination occurs for the loci that are farthest from one another because crossing-over events are more likely when the distance between loci is greater. Therefore, choice D is the best choice. Assume that this map shows only a portion of the chromosome.

26. **D** Note that two options are available for the D gene (D and d), and two options for the F gene (F and f). Therefore, all the gametes have C, E, and g. There are therefore $1 \times 2 \times 1 \times 2 \times 1 = 4$ gametes.

27. **D** Menstruation generally occurs between days 1 and 5 of the menstrual cycle. Ovulation occurs about day 14.

28. **C** Note that the trait is passed from males to females and back to males. In addition, note that the trait is probably dominant because it is represented in each generation. Dominant traits cannot exist in a carrier state. Because sex does not appear to play a role here, the mode of inheritance is most likely autosomal dominant.

29. **D** Use H–W equilibrium. Suppose that p is the frequency of the dominant allele (blue eyes). Because q = 0.2 and p + q = 1, p = 0.8. The frequency of heterozygotes is 2 pq; therefore, $(2)(0.8)(0.2) = 0.32$.

30. **B** A color-blind girl is X^cX^c. Because a father donates an X chromosome to his daughter, this particular father must donate an X^c to have a color-blind daughter.

HIGH-YIELD REVIEW QUESTIONS

Section III: Genetics, Evolution, and Botany

Question Set 9

1. Which factor would tend to shift a population out of Hardy–Weinberg equilibrium?
   A. Barriers to migration
   B. Prevention of mutation
   C. Population-size increases
   D. Preferential mating

2. In the modern understanding of the concept of natural selection, the fittest individuals are those who:
   A. produce the largest number of progeny.
   B. are adapted to the widest diversity of environments.
   C. produce the most highly variable offspring.
   D. have the largest number of fertile offspring.

3. Hemophilia is a sex-linked disease characterized by the inability of blood to clot normally. Consider a famous British royal family that was stricken with hemophilia. A member of this family was Prince Frederick, who was a hemophiliac. Which statement is most likely true of Frederick's family?
   A. His mother was likely to have been a carrier.
   B. His father must have been a hemophiliac.
   C. His grandfather may have been a carrier.
   D. His sister must have been a hemophiliac.

4. A person of blood type A can:
   A. be the parent of a child with blood type B.
   B. possess only the B cellular antigen.
   C. possess both A and B cellular antigens.
   D. safely receive blood from a donor of type B.

5. For sexually reproducing species, which criterion is the most important in determining whether two populations are members of the same species?
   A. Sharing of the same habitat
   B. Mating between the two populations
   C. Morphological similarities
   D. Viable, fertile offspring from both

6. The relationship between two species benefits one species but neither benefits nor harms the other species. This relationship is known as:
   A. mutualism.
   B. symbiosis.
   C. commensalism.
   D. parasitism.

7. The wings on a bird and the wings on a fly are considered:
   A. homologous structures.
   B. analogous structures.
   C. both A and B.
   D. neither A nor B.

8. Consider a population of brown-eyed and blue-eyed people. If brown eyes are dominant over blue, and the frequency of blue-eyed people is 0.16, the fraction of brown-eyed people who are heterozygous is:
   A. 0.48.
   B. 0.57.
   C. 0.66.
   D. 0.42.

9. The A and B blood types exhibit:
   A. complete dominance.
   B. incomplete dominance.
   C. codominance.
   D. no dominance.

10. A brown rabbit mates with a white rabbit and produces all tan rabbits in the F_1 generation. Rabbits of the F_1 generation mate with one another and produce all tan offspring in the F_2 generation. The best explanation for tan rabbits in the F_1 generation is:
    A. complete dominance.
    B. incomplete dominance.
    C. codominance.
    D. no dominance.

11. In cattle, a specific locus on chromosome 9 codes for the horns (dominant), or hornless (recessive) characteristic. A locus on chromosome 6 codes for short hair (dominant) or long hair (recessive). Which ratio is produced from a dihybrid cross of cattle with horns and short hair?

    A. 1 horned with long hair: 1 hornless with short hair
    B. 3 horned with short hair: 1 hornless with long hair
    C. 9 horned with short hair: 1 hornless with short hair
    D. 9 horned with long hair: 1 hornless with short hair

12. Genetic drift refers to:

    A. gene migration and increased variation that upset Hardy–Weinberg equilibrium.
    B. all the genes of a population.
    C. repeated speciation into a variety of ecologic types.
    D. random changes in gene frequency due to small population size.

13. A mature, unfertilized egg cell is found to contain two X chromosomes. This situation is most likely due to:

    A. normal meiosis.
    B. normal mitosis.
    C. nondisjunction in meiosis.
    D. independent assortment in meiosis or mitosis.

14. After several hours, a crow is no longer frightened by a scarecrow. Which type of learning is primarily associated with this change?

    A. Mimicry
    B. Sensitization
    C. Habituation
    D. Imprinting

15. Curly wings in *Drosophila* is due to a rare inherited trait. A curly-winged female mates with a straight-winged male. One half of the offspring produced, male and female, are curly winged. The gene that determines this characteristic is probably:

    A. autosomal dominant.
    B. autosomal recessive.
    C. sex-linked recessive.
    D. two of the above.

16. A trait in a family is investigated. Both parents are heterozygous for the trait. Their offspring demonstrate a 3 to 1 ratio of trait expression to nonexpression. The most likely route of inheritance is:

    A. autosomal dominant.
    B. autosomal recessive.
    C. sex linked.
    D. random mutation.

17. A researcher seeks to discover the order of four genes on a chromosome (A, B, C, and D). The data from numerous recombination events of the four genes on a single chromosome are shown in the following table.

    Recombination Frequencies

	A	B	C	D
A	0	24	42	37
B	24	0	38	10
C	42	38	0	23
D	37	10	23	0

    Based on the data, the order of these genes on the chromosome is most likely:

    A. BCAD.
    B. BACD.
    C. ABDC.
    D. ADBC.

18. The correct taxonomic form is:

    A. homo sapiens.
    B. homo Sapiens.
    C. Homo sapiens.
    D. Homo Sapiens.

19. The terms ranked in increasing order of inclusiveness are:

    A. genus, class, family, order.
    B. species, family, order, class.
    C. order, family, class, phylum.
    D. kingdom, phylum, class, genus.

20. Controlled field experiments with species A, B, and C are performed. Species A is added to an ecosystem containing equilibrium numbers of species A, B, and C. Later, species B and then species C are added under conditions similar to those described for A. A summary of the changes noted in the number of individuals of each species follows:

Species added	Changes
A	B decreases C increases
B	A increases C decreases
C	A increases B increases

    The most likely ecologic relationship of A, B, and C is that:

    A. C preys on A, which preys on B.
    B. A and B are competitors for prey species C.
    C. B preys on C, which preys on A.
    D. A and B are predators; B and C are prey.

21. In a large population, assume that everyone is AA, Aa, or aa with regard to a certain trait: 50% of the AA infants survive past puberty; 75% of the Aa and 5% of the aa individuals do likewise. The term most applicable to this situation is:

    A. linkage.
    B. incomplete penetrance.
    C. incomplete dominance.
    D. heterozygous fitness.

22. An organism has a genotype of XxYyZz. What fraction of its gametes is Xyz or xYZ?

    A. 1/2
    B. 1/4
    C. 1/6
    D. 1/8

23. In humans, brown eyes are dominant over blue. A brown-eyed man, whose parents have blue and brown eyes, respectively, marries a woman with blue eyes. What is the probability of their having four blue-eyed and two brown-eyed children in that order?

    A. 1/1024
    B. 1/128
    C. 1/256
    D. 1/64

24. A man suspects that his wife has had an affair at some point in their marriage. He hires a private investigator to review the situation; yet extensive investigation fails to show any evidence of an affair. Convinced that his wife did have an affair, the man hires a geneticist to review the case.

    The geneticist notes that the man and his wife have three children. He also notes that red–green color blindness, which is inherited as a sex-linked recessive trait, runs in the family. On investigation, the geneticist discovers that the three children are a color-blind son, a son with normal vision, and a color-blind daughter. Both parents have normal vision. Does the man have genetic evidence for his claim?

    A. Yes, because of the color-blind son
    B. Yes, because of the normal son
    C. Yes, because of the color-blind daughter
    D. No, because of no genetic evidence to support his claim

25. Which statement about seeds is NOT true?

    A. Seeds provide enhanced survival of plant embyros.
    B. Seeds contain energy stores for the plant embryo.
    C. Seeds are generally sensitive to desiccation.
    D. All of the above are true.

26. Which plant division contains the largest number of species?

    A. Bryophyta
    B. Sphenophyta
    C. Coniferophyta
    D. Anthophyta

27. Which tissue is found in the center of a plant stem?

    A. Cork cambium
    B. Phloem
    C. Pith
    D. Cortex

28. The reproductive parts of a flower are:
    I. sepals.
    II. stamen.
    III. carpel.
    IV. stomata.

    A. I and II
    B. II and III
    C. I, II, and III
    D. I, II, III, IV

29. Which mechanism does a nonvascular plant use to conduct water through the plant?

    A. Diffusion
    B. Capillary action
    C. Cytoplasmic streaming
    D. All of the above

30. The ultimate source of all genetic variability is:

    A. mutation.
    B. recombination.
    C. crossing-over.
    D. independent assortment.

31. Which structures are contained in plant cells and not in animal cells?

    I. Chloroplasts
    II. Mitochondria
    III. Cell walls
    IV. Endoplasmic reticulum

    A. I and IV
    B. I and III
    C. III and IV
    D. II and III

SOLUTIONS

Genetics, Evolution, and Botany Set 9

1. **D** Hardy–Weinberg equilibrium assumes conditions of genetic equilibrium. Genetic drift, mutations, emigration, and immigration are not allowed in Hardy–Weinberg equilibrium situations, because they shift a population out of genetic equilibrium by changing the genotypes in a population. Random mating must occur in Hardy–Weinberg equilibrium.

2. **D** The key to being the "fittest" is to produce the largest number of fertile offspring. If one produces many offspring and they do not mate, overall one is less successful from a natural selection perspective. Therefore, choice A is not the best choice because it does not specify fertility.

3. **A** Choice A is the best choice. For a male to become a hemophiliac, he must inherit an X^h chromosome from his mother. Since the father donates only a Y chromosome to his son, choices B and C make no sense. If the prince's mother gave him an X^h, she was in all likelihood a carrier. It is very rare for a female to have both X^h chromosomes. His sister can be a hemophiliac only if the father is a hemophiliac and the mother is a carrier or a hemophiliac.

4. **A** If a person with type A (AO) blood mates with a person of type B (BO) blood, their child may have type B blood. A person with type A blood has antigens for type A and antibodies for type B. A person with type A blood can receive blood from type O (universal donor) and can give blood to only type A or AB recipients.

5. **D** Members of the same species should be able to mate and produce viable, fertile offspring. For example, a dog and a cat are different species because they cannot produce viable offspring. A horse and a mule can mate and produce viable offspring (a donkey); however, a donkey is sterile. Therefore, horses and mules are members of different species.

6. **C** Commensalism occurs when the relationship between two species is such that one species benefits and the other neither benefits nor loses. In mutualism, both parties benefit from the association. Symbiosis allows one organism to live in harmony and derive benefit from another. Parasitism refers to a relationship in which one party benefits from the association at the expense of the other party.

7. **B** Analogous structures do not have similar developmental origins, whereas homologous structures do. Examples of homologous structures include bird wings and bat wings.

8. **B** The q^2 term of the H–W formula ($p^2 + 2pq + q^2 = 1$) is 0.16; therefore, $q = 0.4$. Then $p = 0.6$, because $p + q = 1$. The fraction of brown-eyed people who are heterozygous $= 2pq/(p^2 + 2pq) = 2(0.6)(0.4)/\{0.6^2 + 2(0.6)(0.4)\} = 0.57$.

9. **C** Codominance is the answer because neither type A nor type B is dominant over the other. As a matter of fact, patients can be type AB. No "blending" is seen in the phenotype of codominant crosses although blending is seen in cases of incomplete dominance (red flowers crossed with white flowers = pink flowers).

10. **B** In this example of classic incomplete dominance, "blending" occurs in the phenotype. Complete dominance would produce rabbits of only one color. Rabbits would be either brown or white, whichever was the dominant allele. Codominance would not produce a "blending" of rabbit color.

11. **A** One should know the classic 9:3:3:1 ratio of dihybrid crosses. This ratio means that of 16 possible genotypes generated from the cross, 9 exhibit double dominance, 3 show dominance of each of two genotypes, and one shows double recessive genotype. For this question, 9/16 show horns and short hair, 3/16 show horns and long hair, 3/16 show no horns and short hair, and 1/16 shows no horns and long hair. Choice A is correct because the ratio of horns/long hair to hornless/short hair is 3:3, which reduces to 1:1.

12. **D** Genetic drift refers to random changes in gene frequency due to a small population size. For example, consider a population of five people. If one person in this population has red hair and is the only person to have a gene for red hair color,

what happens to the gene frequencies if something happens to this person? If a random event (such as a bolt of lightening) kills this person, the gene frequencies for hair color in this population would change dramatically. This example shows how genetic drift can affect the gene frequencies in small populations.

13. **C** A mature, unfertilized egg cell should contain only one sex chromosome. Because the sperm contributes a sex chromosome, having two sex chromosomes in the nonfertilized egg increases the risk of birth defects. The most likely mechanism for this problem is nondisjunction during meiosis.

14. **C** Habituation is seen when a stimulus is repeated many times and the organism begins to react differently to the stimulus because of familiarity. Mimicry occurs when one species benefits by resembling a different species. For example, a butterfly may have a pattern of a large eye on its wings to look like a larger predator. This pattern may help protect the butterfly from its predators. Sensitization occurs when a repeated stimulus increases the ultimate response to the stimulus. Habituation is a type of learning that involves a loss of responsiveness to unimportant stimuli. For example, a mouse may no longer become alerted to a loud noise when the noise is repetitive. Imprinting is irreversible learning that occurs at a critical time in development. For example, ducklings that hatch and spend their first few hours of life in the presence of a human may believe that the human is their mother.

15. **A** Possessing curly wings in the first generation appears independent of sex. In addition, one can expect to see carriers among females if sex-linked recessive is the form of transmission. Therefore, a dominant form of transmission has occurred. Autosomal dominant transmission is plausible.

16. **B** Recessive disorders skip generations because heterozygotes are carriers and are phenotypically normal. The trait is autosomal because the sex of the parents and offspring is unimportant.

17. **C** The farther apart two loci are on a chromosome, the greater is their recombination frequency. Therefore, A and C have the highest frequency (42), whereas B and D have the lowest (10). One can surmise that A and C are the farthest apart and B and D are the closest. Because A and B are closer than A and D, the best answer is choice C.

18. **C** The genus name should be capitalized and underlined. The species name should not be capitalized, but it should be underlined.

19. **B** The correct order of decreasing inclusiveness is kingdom, phylum, class, order, family, genus, species. Therefore, choice B is a correct list of increasing inclusiveness.

20. **D** For this difficult question, use the process of elimination. Notice that the addition of species C increases A and B. C is likely a foodstuff or prey for either A or B. Note that when A is added, B decreases and C increases. This pattern suggests that A is a predator for B and that the decrease in B allows the number of C organisms to increase; this situation is true if B is a predator for C.

21. **D** The heterozygotes fare better than the homozygotes. This situation is known as heterozygous fitness. Linkage usually describes a situation in which a particular gene is linked to a specific chromosome. An example would be sex-linked traits such as color blindness. The gene for color blindness is linked to the X sex chromosome. Incomplete penetrance refers to a situation in which a genotype does not always give the predicted phenotype. For example, certain autosomal dominant diseases can be expected to affect 1/2 of a family of afflicted individuals. This disease is seen in 1/2 of the family, but different family members with the disease have different levels of involvement or disease activity. Some family members with the disease gene are very disabled, whereas others show hardly any signs of the disease.

22. **B** The fraction of zygotes Xyz = $(1/2)^3$. The fraction of zygotes xYZ = $(1/2)^3$. The fraction of Xyz or xYZ = 1/8 + 1/8 = 1/4.

23. **D** The man is heterozygous, Bb. The wife is homozygous recessive, bb. The cross Bb × bb gives 1/2 offspring with brown eyes and 1/2 with blue eyes. The chance of four blue and two brown in order is $(1/2)^6$ or 1/64.

24. **C** To have a color-blind daughter, both parents must have an Xc gene. Because the male parent has normal vision, and men cannot be carriers of X-linked traits, the man's wife has had an extramarital relationship.

25. **C** Choices A and B are true statements. Seeds provide a significant advance to seed plants because seeds protect the embryo and contain energy stores for the embryo. Choice C is an incorrect statement because seeds protect the plant embyro from desiccation. Evolution of the seed makes it possible for plants to reproduce in nonmoist environments.

26. **D** The angiosperms (Division Anthophyta) represent the largest division of species, that is, approximately 250,000 different species. The other divisions listed include Bryophyta (mosses), Sphenophyta (horsetails), and Coniferophyta (conifers). These divisions all contain far fewer species.

27. **C** The pith is tissue found in the center of a stem or a trunk. It acts to store food for the plant. The cork cambium is just inside the outer cork layer of a tree stem. It is clearly more external than the pith layer, and its role is to produce new cork cells. The cortex is the first layer inside the cork cambium. It is also external to the pith and serves a food storage function. The phloem is found inside the cortex layer, but external to the pith. The phloem is involved in food transport through the stem or trunk.

28. **B** The reproductive parts of a flower are the stamen and carpel. The stamen produces pollen grains (male gametophytes). The carpel contains the ovary. The ovary has one or more ovules. An egg cell contains a female gametophyte. Sepals serve to protect the floral bud before it opens. Sepals are usually thickened green structures that wrap over a closed floral bud (e.g., a rosebud). Stomata are tiny pores in leaves of plants; they penetrate the epidermis layer of a leaf to allow gas exchange with the atmosphere.

29. **D** Because nonvascular plants do not contain tubes (xylem) to conduct water within the plant, they use diffusion, capillary action, and cytoplasmic streaming to distribute water throughout the plant.

30. **A** Only mutations introduce new DNA structure to the genome. Recombination and crossing-over just rearrange existing genetic loci.

31. **B** Plant cells differ from animal cells in having several unique structures. Plant cells contain chloroplasts, a central vacuole, a cell wall, and plasmodesmata. Both animal and plant cells contain mitochondria and endoplasmic reticulum.